T0135607

Development of the CEMAX system for cell line development

based on site-specific integration of expression cassettes

Dissertation
zur Erlangung des akademischen Grades
des Doktors der Naturwissenschaften (Dr. rer. nat.)

an der Technischen Fakultät
der Universität Bielefeld

vorgelegt von

Benedikt Greulich

geboren am 23. Mai 1980 in Warendorf

November 2010

Gedruckt mit Genehmigung der Technischen Fakultät der Universität Bielefeld.

Dissertation, Technische Fakultät der Universität Bielefeld

Betreuer und Erstgutachter: Prof. Dr. Thomas Noll
Zweitgutachter: Prof. Dr. Hermann Ragg

Tag der mündlichen Prüfung: 14.03.2011

Bibliografische Information der Deutschen Nationalbibliothek

Die Deutsche Nationalbibliothek verzeichnet diese Publikation in der
Deutschen Nationalbibliografie; detaillierte bibliografische Daten sind
im Internet über http://dnb.d-nb.de abrufbar.

ISBN 978-3-8325-3112-6
ISSN 1866-9727

Logos Verlag Berlin GmbH
Comeniushof, Gubener Str. 47,
10243 Berlin
Tel.: +49 (0)30 42 85 10 90
Fax: +49 (0)30 42 85 10 92
INTERNET: http://www.logos-verlag.de

ABSTRACT

The random nature of transgene integration requires an intensive clone screening process for the establishment of a production cell employing high yield and quality of biopharmaceutical, recombinant proteins. This thesis was focused on the development of an expression system that makes use of site-specific integration of the product gene in CHO-K1 cells combined with serum-free cultivation. Common barriers in protein production from stable cell lines should thus be overcome. The goal was achieved by the development of the CEMAX system that allows site-specific integration of expression cassettes at highly active sites of transcription. Producer cells were generated using a DNA double-strand break induced homologous recombination mechanism based on recombinant CEMAX host cells. This enabled the rapid and reproducible establishment of stable producer cell lines with known characteristics. Productivities of up to 10 pg per cell and day could be achieved even for the production of highly glycosylated proteins. In a separate aspect of this work CHO-K1 host cells growing in suspension in serum-free medium were improved to reach viable cell densities of 10^7 cells per ml an allow cloning efficiencies up to 71% in serum-free medium.

Acknowledgements

A lot of people contributed to this work in a plenty of ways. It would not have been possible this way without all the encouragement.

First, I am pleased to thank Prof. Dr. Thomas Noll for his readiness to be my advisor at the university. He provided help whenever necessary and great support during the specification of supplementary studies.

Dr. Andreas Herrmann has made this work possible in a number of ways. He provided the possibility to work on an interesting and challenging PhD project in an industrial environment. I am deeply grateful for all his help and guidance during the thesis and his useful comments in scientific discussions and on the manuscript.

I would like to thank Prof. Dr. Karl Köhrer and Dr. Sebastian Wohlfromm from the BMFZ of Heinrich Heine University Düsseldorf for a fruitful collaboration and their work on characterisation of cell lines at the molecular level including Southern blot analysis.

Prof. Dr. T. Scheper and PD Dr. C. Kasper gave me the possibility to perform fluorescence-activated cell sorting at the TCI of Leibniz University Hannover. In this regard I would like to thank especially Pierre Moretti for his great commitment during the collaboration, although the work on the third series of host cells did not provide direct improvements of the CEMAX system. Furthermore, I would like to thank Prof. Dr. R. Küppers and K. Lennartz from the IFZ of University Duisburg-Essen for fluorescence-activated cell sorting during work on the fourth series of host cells.

I am pleased to thank my former colleagues in Jülich for nearly five valuable years during internship, diploma thesis and the first part of this work. In particular I would like to mention Silke Schindler and Melanie Rüping for their technical assistance, and Verena Lorenz for providing the data on Atrosab expression. Anja Urbschat assisted during screening of host cell candidates and optimisation of cotransfection parameters in the course of her diploma thesis, which was integrated in this work.

I am grateful to my colleagues in Basel for a pleasant working environment during the last year of the thesis. Especially Dr. Karlheinz Landauer raised some interesting questions during scientific discussions and was a great help by reviewing the manuscript.

I thank Anne-Kathrin John for helpful comments to the method section and the introduction.

Great thanks also go to Prof. Dr. J. Müthing and his group for the organisation of two outstanding weeks in Münster in July 2006.

I thank my brother Johannes for the discussion about the theoretical considerations on limited dilution cloning and his support with the mathematical operations. Last but not least, I would like to thank my parents for all their encouragement through the last years, and especially Maria and recently Anna Lena and Pia Luisa for giving me some extra motivation and a great experience.

Table of Contents

1 Introduction

Since the first use of recombinant DNA technology by Cohen and colleagues in 1973 [1] and the approval of recombinant human insulin in 1982 as the first biopharmaceutical for treatment of human disease, over 165 recombinant products have been approved for human use until 2006 [2]. The absence of complex posttranslational modifications of insulin allowed its large-scale production in a microbial process using *Escherichia coli* as host. However, mammalian cells, which are capable of mimicking characteristic modifications found in human proteins, are the only choice for recombinant production of complex proteins that require proper assembly, folding and posttranslational modifications available so far. Thus, more than twenty years after the approval of the first recombinant proteins from cell culture processes, which were tissue plasminogen activator in 1986 and erythropoietin in 1989, mammalian cells are the predominant expression system for complex recombinant proteins. Today about 60% to 70% of recombinant protein biopharmaceuticals are delivered by mammalian cell culture processes [3]. To cope this evident demand the world wide cell culture capacity has increased to estimated 475000 litres in 2005 [4]. On the other hand, steadily increasing yields surpassed product concentrations of 5 $g \cdot l^{-1}$, representing a more than 100-fold improvement compared to values obtained using adherent cells in the mid-1980s [3].

Today's strategies in cell line development suffer from one main limitation: high producers have to be identified in a screening process that is intensive in terms of labour and costs. This is mainly due to the random process of transgene integration with variable copy numbers and position effects that causes a wide range of transcription levels and thus dictates the time consuming screening process.

The typical development for investigational new biological drugs takes between 8 and 15 years after start of preclinical work. It is associated with high costs and low rate of approval [5, 6]. Taking into account the patent expiry after 20 years, every month faster in development of a blockbuster drug (> \$ 1 billion sales per year) is equivalent to at least 80 million € revenue.

Gene targeting would supersede clone screening approaches in the field of recombinant protein production from stable cells if the target locus allows high transcription rates. This would reduce timelines in drug development by 3 to 6 months. In addition, other applications that profit from the reproducibility effect of gene targeting would be potential fields for implementation. Clones generated by gene targeting should be similar to the host cell clone regarding growth characteristics, integration site, and copy number. Therefore they should provide similar expression levels and product quality. For example in cell line engineering, targeted integration allows to study the influence of particular factors expressed in a cell without coping with clonal differences.

DNA double-strand break induced homologous recombination is an established tool for genome engineering, but has not been adapted to the field of recombinant protein production.

2 Objective

The objective of this work was to develop a technology for the development of producer cells for the production of biopharmaceutical proteins by the use of targeted integration in order to make the cell line development process faster, more convenient and reproducible.

In order to achieve site-specific integration, DNA double-strand break induced homologous recombination technology has been chosen. Serum-free adapted CHO-K1 cells as host and homing endonuclease I-SceI for delivery of DNA double-strand breaks should be used in the development that would include the following steps:

1. Establishment of a concept and development of a vector system for *in vivo* recombination and selection of targeted cells

2. Cloning of tag and replacement vectors

3. Generation of modified host cells that provide high expression levels; assessment of productivity with a reporter gene

4. Targeted integration of the gene for green fluorescent protein as a proof of concept; establishment of targeting and selection protocols

5. Generation and characterisation of producer cells for biopharmaceutical product candidates

3 Literature survey

Although various strategies exist for both cell line development and gene targeting a combination of these in the field of recombinant protein production is still not well established. The first part of this literature survey is focused on production of recombinant proteins from transfected mammalian cells and strategies for the development of production cell lines. Gene targeting and mechanisms of recombination are included in the second part, which also deals with the homing endonuclease I-SceI that can be used to improve the frequency of gene targeting.

3.1 Recombinant protein production in mammalian cells

The expression of a recombinant protein necessarily begins with the protein-coding gene of interest (GOI). In times of synthetic genes the coding sequence can be optimised easily in terms of codon usage, elimination of cryptic splice sites, and RNA destabilising motifs to improve recombinant protein production [7]. A Kozak sequence [8, 9] is usually introduced flanking the start codon of the product gene to optimise initiation of translation. The product gene is cloned into a recombinant expression vector usually containing a strong cellular promoter, for example from the elongation factor-1α gene [10-12], or a viral promoter for initiation of transcription. One of the strongest and most widely used viral promoters is derived of the regulatory sequences of the immediate early gene of human or murine Cytomegalovirus (CMV) [13, 14]. Highest expression rates are achieved in combination with an intron that has been shown to enhance productivity [12, 15]. The expression cassette is completed with a 3' untranslated region that includes a cellular or viral polyadenylation signal [16, 17] for termination of transcription and increased mRNA stability. For selection of stable cell lines the plasmid vector contains a selectable marker such as for instance neomycin phosphotransferase (neo) [18], hygromycin phosphotransferase (hygro) [19], or a selection marker allowing gene amplification. Recent approaches tend to use selectable markers with reduced expression rate and physical linkage to the product gene [20], markers with reduced protein activity [21] or trans-complementing selection markers [22] to increase the expression level of stable clones. Most popular selection markers for gene amplification strategies are the dihydrofolate reductase (DHFR) [23, 24] and the glutamine synthetase (GS) [25]. These markers allow selection of stable cells in the absence of a specific metabolite by preventing growth of untransfected cells. Gene amplification is then achieved by inhibition of the marker with methotrexate (MTX) or methionine sulfoximine (MSX), respectively. The DHFR system is based on Chinese hamster ovary (CHO) cell mutants deficient in DHFR activity due to gene deletion and inactivation [26, 27], whereas the GS CHO system does not require mutant cells along with other advantages like less production of ammonia due to intracellular production of glutamine and faster development times [28]. Further vector elements, such as scaffold- or matrix-attachment regions (S/MAR) [29-31], isolators [32], stabilising and anti repressor (STAR) elements [33], expression augmenting sequence elements (EASE) [34], and ubiquitous chromatin opening elements (UCOE) [35] have demonstrated potential to increase productivity of stable cell lines. The effect of these sequence elements, of which some were isolated from genomic sequences surrounding highly expressed genes, is mainly based on compensation of negative position effects. Nevertheless, neither gene amplification nor sequence elements contribute much to reduced development times in cell line generation, since these methods are still dependent of an intensive clone screening process.

Once the product gene is cloned into an appropriate expression vector, it is transfected into the cell line chosen for production. CHO cells have become the standard mammalian host cell for protein production [4, 36]. But other cell lines, for example derived from baby hamster kidney (BHK), mouse myeloma (NS0 and SP2/0), human embryonic kidney (HEK 293) and human retina (PER.C6®) are also used for recombinant protein production. A number of possible methods for gene delivery are available including calcium phosphate precipitation [37], electroporation [38], lipofection [39] and polymer-mediated gene transfer, for example via polyethylenimine (PEI) [40]. DNA uptake and incorporation into the genome gives transfected cells a selective advantage to survive antibiotic selection, which is initiated one or several days after transfection. Survivors are expanded from single cells to give rise to clonal populations, which are characterised for their productivity of the recombinant protein. The identification of high producers is a tedious and time-consuming screening step and requires assessment of hundreds to thousands of clones, since specific productivity of cells spans a wide range. Based on positional effects and variances in transgene copy numbers [41] caused by random integration of the transgene combined with different secretory properties of individual clones [42] high producers occur rarely. The screening process is usually performed by limited dilution cloning in a serum-containing environment to reach cloning efficiencies that make it applicable. The process of single cell cloning can be streamlined by cost-intensive equipment like cell sorters, clone pickers and robotic systems [43] but this is usually only implemented at large companies, which can afford high investments. After this initial screening round a set of highest producers is analysed for characteristics including productivity, stability of protein expression, growth characteristics and biochemical protein quality making them good production clones. Owing to instability of protein production based on gene silencing and genomic instability several clones have to be excluded from further experiments, which is a particular drawback of gene amplified clones [44, 45]. A second round of limited dilution cloning is performed to ensure clonality of the production cell line. Out of these candidates an appropriate clone is chosen for development of a production process. The production capacity of these high-yield cell varies between 15 and 60 pg·cell^{-1}·day^{-1} (pg·c^{-1}·d^{-1}) [43] and can reach values of up to 100 pg·c^{-1}·d^{-1} for antibody products [46]. However, these values are usually not achieved with other classes of recombinant proteins. Productivities >10 pg·c^{-1}·d^{-1} are achieved less commonly with proteins for which posttranslational processing and especially glycosylation might be the limiting step [46].

Although these methods are capable of delivering highly productive clones, they require gene amplification accompanied with genetic instability and screening of usually above 1000 individual clones what results in cell line development times of six to nine months. Targeted integration of the product gene into a genomic locus providing high transcription rates would supersede the screening process and extensive clone characterisation and thus reduce the development times by at least three months. However, there are no reports of cell line development strategies using targeted integration reaching the productivities described above.

Production of therapeutic protein candidates for *in vitro* or *in vivo* characterisation in early stages of drug development usually requires less protein material and therefore less productive processes are sufficient. More important than productivity in this case is that expression is fast and efficient. These proteins can be either expressed transiently, although product concentration is usually below 100 mg·l⁻¹ [47] and more or less in the range of 10 mg·l⁻¹ [48, 49], or with stable cell lines, which is usually time consuming. Targeted integration would be of particular interest in this field of expression since it does not depend on a screening process and is thus fast. A second virtue is that protein isoforms would be expressed in the same cellular background: the integration site is identical as well as other clonal characteristics including growth characteristics and the potential of posttranslational modification. Proteins expressed from one host cell clone would therefore only vary in amino acid based differences.

Process development of animal cell cultures is an iterative and labour intensive technique. Optimisation of cultivation strategies is usually performed in scale-down models. Most widely used are bioreactors in the 1 L scale [3] and non instrumented screening tools with working volumes below 100 ml. Among small-scale cultivation systems disposables are replacing glassware roller bottles and spinner flasks. The most widely used small-scale systems now include well plates, T-flasks, and shake flasks. Shaking technology and centrifugation tubes adapted for cell cultivation, for example, was used successful for evaluation of a number of process parameters [50]. Among reactor types for large-scale manufacturing the stirred tank reactor (STR) made of stainless steel is the reactor type of choice. Nevertheless, disposable bioreactors (e.g. Wave, CELL-tainer, S.U.B.) are used for inoculation train and production of material for clinical phases in working volumes up to 1000 litres.

One of the main driving factors for improved process yields was the development of serum-free media in the past two decades. The use of serum in early cell culture processes resulted in substantial scale-up challenges [46] and raised regulatory concerns. Today chemically defined media allowing suspension cultivation at high cell densities and high viabilities, and are available off-the-shelf. Nevertheless, large-scale manufacturers invest substantial effort in development of proprietary media formulations optimised for the metabolic demand of clonal production cell lines [3, 51].

Besides optimised culture media, the mode of cultivation contributed to improvements in process yields in the past decades. Peak cell densities of $3*10^6$ cells per ml and product concentrations around 100 mg$\cdot$L^{-1} were achieved by batch cultivation in processes more than twenty years ago [36]. Today fed-batch cultivation is most widely used in production processes due to scalability, ease of operation and high volumetric productivity. Viable cell concentrations exceed $10*10^6$ cells per ml and product concentrations in the range of 1-5 g$\cdot$L^{-1} for antibodies are achievable [3, 36]. Continuous perfusion processes allow cultivation at even higher cell densities due to cell retention and continuous medium replacement in periods of several months. This strategy was, for example, applied for the production of the fragile recombinant factor VIII [52].

Typical fed-batch development strategies include basal medium optimisation or screening of commercially available media. This step is followed by feed medium optimisation and development of supplementation strategies to meet clonal nutrient requirements. Optimal process parameters like pH and dissolved oxygen level are evaluated employing controlled scale-down models of bioreactors. The scale-up to production scale is then performed stepwise during drug evaluation phases as reported, for example, in a recent case study of a mouse myeloma process [53].

3.2 Gene targeting

Various methods are available for stable transfection of eukaryotic cells, but the incorporation of heterologous sequences into the genome is usually not under control. Gene targeting, the recombination of a transfected DNA molecule with a specific chromosomal locus, is a mean to control the integration site of heterologous sequences. Gene targeting by homologous recombination between an artificial vector and an endogenous locus was first reported by Smithies *et al.* in 1985 [54]. In the beginning, selection of targeted events was performed by positive selection with markers such as neomycin phosphotransferase (neo) combined with screening for correctly targeted clones. Depending on the extend of homology, targeted clones occurred at low frequency of one positive event in 1000 to 40000 Geneticin (G418) resistant cells [54, 55].

3.2.1 Classification of targeting vectors

As classified by Thomas and Capecchi, two types of targeting vectors can be distinguished: replacement vectors and insertion vectors [55], also referred to as ends-out and ends-in targeting, respectively [56]. The two types of targeting vectors are exemplified for the targeting of the hypoxanthine phosphoribosyl transferase (hprt) gene in mouse embryonic stem cells (ES cells) in Figure 1. The replacement vector is collinear to the target site while the insertion vector is linearised within the region of homology. The major difference between these targeting vectors is the fact that, upon recombination, the complete vector sequence of the insertion vector is integrated at the target locus whereas only a mutated stretch of sequence is changed with the replacement vector. Higher accuracy of recombination is another advantage of the replacement vector. Targeted integrants could be characterised exclusively as a result of a replacement reaction when using the replacement vector, whereas 25% of recombination events with the insertion vector were characterised by replacement instead of the expected insertion recombination.

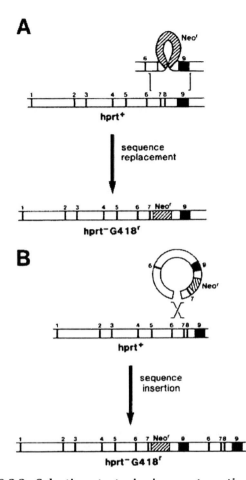

Figure 1: Classification of targeting vectors.

A: Replacement vector as originally described by Thomas and Capecchi [55], which is aligned colinear to the target locus when the vector is linearised. In this case exon 8 is replaced by the neo gene. Beside random integration events no further vector sequences become integrated. B: The insertion vector was used by Smithies *et al.* in the first targeting experiment of an endogenous locus [54]. After linearisation the ends of insertion vector lie adjacent to another. Following recombination with the genomic homolog, the complete vector is integrated resulting in a partial duplication of the target locus. From Thomas and Capecchi [55].

3.2.2 Selection strategies in gene targeting

Significant improvements in the selection strategy of targeted clones were achieved with a positive-negative selection strategy developed in the laboratory of Capecchi. This strategy, which is illustrated in Figure 2, includes elimination of random integrants by negative selection and allows an enrichment of correctly targeted recombination events up to 2000-fold [57]. This is achieved with the negative selection marker thymidine kinase (HSV-tk) from Herpes simplex virus (HSV) that is lost during homologous recombination in targeted cells but not during random integration. The nucleoside analogue ganciclovir is phosphorylated by HSV-tk and then potently disrupts DNA synthesis and thus kills cells that have undergone random integration.

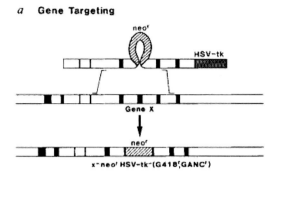

a **Gene Targeting**

b **Random Integration**

Figure 2: Gene targeting using positive-negative selection strategy.

Exon 2 of the target gene X is disrupted by the positive selection marker neo in a replacement vector. The negative selection marker thymidine kinase (HSV-tk) from Herpes simplex virus (HSV) is lost during homologous recombination in targeted cells (a) and is used to eliminate random integrants that commonly contain an intact copy of the HSV-tk gene (b). From Mansour *et al.* [57].

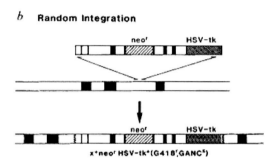

An alternative selection procedure to target genes that are expressed in the cell type used for the experiment is conditional positive selection. A regulatory element that is not included in the selectable marker is provided by the target gene and thus allows proper expression of the marker dependent on homologous recombination. An example for this strategy is a neo expression cassette lacking a polyadenylation signal, which is provided by downstream sequences of the target gene allowing expression of the marker after homologous recombination [58].

An interesting variant of conditional selection is a two-step strategy termed "plug and socket" developed by Detloff *et al.* [59]. This strategy comprises a modified endogenous locus for repeated targeting by direct positive selection (Figure 3). Selection is based on the completion of two deletion mutants of a positive selection marker. In a first step, a 5' deletion mutant (ΔM2) is introduced in an artificial target site near the gene X that should be altered. The desired mutation is introduced into gene X in the second step. This is done via homologous recombination at endogenous sequences (left hand) and overlapping sequences between ΔM2 and a second deletion mutant of M2 (M2Δ) lacking the 3' part of the gene (right hand). Recombination between the two deletion mutants delivers a functional selection marker (M2) allowing direct positive selection of targeted cells. This is especially profitable when several mutation variants of gene X should be generated. The strategy has been applied with the hprt marker [59, 60] and was adopted for the use with hygro [61], for example.

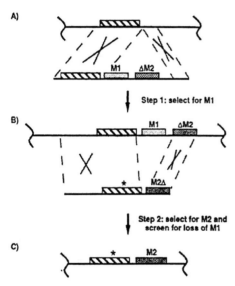

Figure 3: Plug and socket strategy for repeated gene targeting.
The locus containing the gene X to be altered (cross-hatched box) is engineered with a socket targeting construct containing a functional selectable marker gene (M1) and a second nonfunctional marker gene (ΔM2) via homologous recombination and selection for M1 (A). The real targeting experiment is then performed with a plug targeting vector and involves the completion of M2 after homologous recombination with the socket (B, C). From Detloff et al. [59].

3.2.3 Frequencies of gene targeting events

The absolute frequencies for gene targeting events seem to be locus dependent, what may be somewhat due to hindered accessibility of the enzymatic recombination machinery at given loci due to differences in chromatin structure [57, 62]. Targeting frequencies at the hprt locus in mouse ES cells with targeting vectors containing about 2 to 4 kb homology are in the range of $2 \cdot 10^{-8}$ to $2 \cdot 10^{-7}$ events per treated cell as reviewed by Capecchi [63]. The villin locus in mouse ES cells, for example, seems to be less accessible for gene targeting. A frequency of $5 \cdot 10^{-8}$ was achieved with a total of 9.6 kb homology to the target site. Fitting into the range of these findings is the frequency at the adenine phosphoribosyltransferase (aprt) locus in a CHO cell mutant. $4 \cdot 10^{-7}$ events per treated cell were achieved with a total degree of homology of 2.6 kb [64].

Factors that affect gene targeting frequencies positively include the routine use of optimised selection protocols and linearised targeting vectors. Furthermore the use of isogenic DNA sequences [65] and incorporation of elongated stretches of homologous sequences [55, 66, 67] has proven to support recombination.

3.2.4 Improvements in gene targeting and use in permanent cell lines

Although gene targeting by homologous recombination is a powerful tool used for ES cell transgenesis, its potential in permanent cell lines is limited by the fact that homologous recombination events are masked by massive illegitimate recombination events [68]. In other cell systems targeted integration of a transgene has thus to be enhanced or supported by additional artificial introduced enzymatic means for instance.

Observations in experiments with yeast established the recombinogenic nature of DNA double-strand breaks (DSBs) [69]. On the one hand, DSBs have subsequently found acceptance as a prerequisite for homologous recombination in several recombination models (see chapter 3.4). On the other hand, site specific introduction of a DSB by the rare-cutting homing endonuclease I-SceI has proven to induce homologous recombination on extrachromosomal substrates in plant cells [70] and mammalian cells [71]. The potential of this enzyme prototype to enhance the frequency of gene targeting events has been shown after these initial experiments with extrachromosomal substrates. Gene targeting frequency could be enhanced up to 2 to 3 orders of magnitude by specifically introducing a DSB at the target site of a modified chromosomal locus [61, 72-74]. DSB induced homologous recombination in two-step strategies for repeated targeting of a specific locus can be used similar to or in combination with the "plug and socket" strategy. In the first step a recognition site for the homing endonuclease and selection markers are integrated near the gene, which should be mutated in subsequent steps, via classical gene targeting using homologous recombination. This modified locus can then be targeted repeatedly at high frequencies via DSB induced homologous recombination to introduce the desired mutations. Cohen-Tannoudji *et al.* observed a gene targeting frequency at the villin locus in mouse ES cells of 6∙10⁻⁶ events per treated cell, which is approximately a 100-fold increase in frequency by concurrent 3.6-fold reduction of the extend of homology [61].

Site specific recombinases (Cre from bacteriophage P1 and FLP from *Saccharomyces cerevisiae* among others) are a second tool allowing targeted modification of genomes, which has been demonstrated by Sauer and Henderson [75] and O'Gorman *et al.* [76], respectively. These recombinases catalyse a reciprocal, site-specific recombination between two identical target sites of 34 bp and 48 bp, respectively [77, 78]. Thus, recombinases have target requirements on both the integration site and the extrachromosomal recombination partner and at least one target site remains at the altered locus [77]. This makes integration a reversible process in that excision of the intervening sequence is catalysed between two target sites, which is in contrast to the scare less and irreversible process of DSB induced homologous recombination [79].

Although gene targeting has proven to be a vital mean for introducing targeted changes into the mouse genome and generation of transgenic animals, it was not used for the generation of production cells for biopharmaceutical products so far.

3.3 I-SceI and intron homing

I-SceI, the intron-encoded endonuclease I from *S. cerevisiae*, is the first known representative of intron encoded proteins now known as homing endonucleases [80]. The rare cutting enzyme belongs to the LAGLIDADG family of homing endonucleases [81] and appears as a monomeric globular protein with an apparent molecular weight of 26 kD when expressed in *E. coli* [82, 83]. Acting as a monomer containing two LAGLIDADG motifs, I-SceI thus recognises a non-symmetrical recognition site that extends over 18 bp and leaves 4 bp staggered DNA ends with phosphoryl groups at the 5` recessed end [84]. Formerly known as *omega* transposase, I-SceI is encoded by an open reading frame in an optional group I intron in the mitochondrial 21S ribosomal RNA gene of *S. cerevisiae* ω⁺ strain [80, 85]. This mobile intron is transmitted through a population via intron homing (Figure 4). Intron homing is initiated directly after mating of a ω⁺ strain and an intronless ω⁻ strain by I-SceI catalysed cleavage at the junction of the two exons in the intronless allele where the intron is inserted via subsequent gene conversion. The open reading frame encoding I-SceI is thereby essential for intron mobility [86, 87] since a DSB at the target site initiates the process of intron homing.

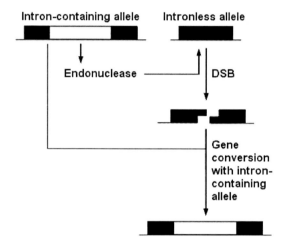

Figure 4: The mechanism of intron homing.

An intron is copied from an intron containing allele to the intron less allele via gene conversion. This process is initiated by a DNA double-strand break introduced into the intronless allele by a homing endonuclease which is itself encoded by the mobile intron. Adapted from Jasin [79].

In vitro I-SceI itself is sufficient to recognise its specific cleavage site and to catalyse DSBs [82, 83]. The 18 bp cleavage site of I-SceI is expected to be found once in $6.9 \cdot 10^{10}$ bp of random sequence and thus probably not in approximately $3 \cdot 10^9$ bp of the Chinese hamster genome. Heterologous expression of active endonuclease with the universal code equivalent of the mitochondrial I-SceI gene [82] has been shown in various organisms and cell systems ranging from *E.coli* and yeast to plant and mammalian cells [61, 70-74, 82, 88, 89], including CHO cells [90]. Expression of I-SceI seems not to be toxic in most of the cell types studied so far. Mouse 3T3 cells with stable expression of I-SceI showed apparently normal growth characteristics [71]. This observation underlines a striking virtue of I-SceI compared to other homing endonucleases like I-CreI and I-PpoI, which proved to be toxic to human cells due to cleavage sites in ribosomal DNA arrays [79].

The capability to introduce one or few DNA DSBs in complex genomes has made homing endonucleases in general and especially I-SceI useful tools for studying DSB repair pathways, cloning and gene targeting experiments.

3.4 DNA double-strand break repair and homologous recombination

DNA is under constant attack from endogenous and exogenous sources. An accumulation of DNA damage can lead to severely impaired cellular functions or might trigger cell death via apoptosis. Although mutations are essential for evolution, genetic stability is important for short-term survival of an organism. Therefore, a network of DNA-repair mechanisms ensures genome integrity. Depurination, desamination or covalent linkage of adjacent pyrimidine bases, for example, affect only one strand of the DNA double helix. As a result, this damage can be repaired efficiently in pathways that use the complementary strand of the double helix to restore the genetic information. These repair mechanisms include the base excision repair and the nucleotide excision repair pathways. Particular genotoxic forms of DNA damage represent DSBs, which affect both strands of the DNA helix and can be inducers of severe threats through chromosomal aberrations [91]. One of the most prevalent sources of DSBs is replication, where single strand gaps in the template and collapsed or broken replication forks may cause DSBs. Besides, these breaks can originate from topoisomerase and endonuclease cleavage, V(D)J-recombination as well as somatic hypermutation for instance. In addition to cellular processes DSBs can arise by ionising radiation, DNA damaging agents and mechanical stress. Protection from these harms is ensured via several pathways in the cellular DSB repair repertoire, which are either based on homologous or illegitimate recombination. These pathways differ in their requirements of homologous sequences as well as in the outcome of the repair reaction.

Non-homologous end joining (NHEJ) is a repair process that uses illegitimate recombination and thus does not require a homologous template. The reaction is outlined in Figure 5 and seems to be prevalent in mammalian cells as a fast means to remove a potential lethal damage [92, 93]. Upon lesion, DNA ends are processed to become compatible and are then reconnected by ligase activity from the repair complex. Ligation of compatible or blunt ends will result in accurate repair whereas deletions and insertions may be introduced if the ends are incompatible.

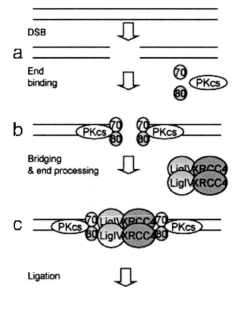

Figure 5: DSB repair through NHEJ. A DSB (a) is bound by a Ku70/80-DNA-PK complex, which might also be involved in initial recognition of DSBs and structurally support during rejoining (b). Processing of the DNA-ends results in removal or addition of a few base pairs depending on the grade of compatibility of the DNA ends. DNA ligase IV in complex with XRCC4 joins the repair complex (c) and catalyses the joining of the ends (d). Reviewed by Helleday [92] and Pfeiffer *et al.* [93]; figure modified from Pastwa and Blasiak [94].

The simple pasting of DNA ends during NHEJ can be accurate if compatible or blunt ends are ligated, but leaves small deletions and insertions at the break site in most cases. In a study of Jasin and colleagues insertions of up to 200 base pairs (bp) beside deletions in the range of one to 21 bp have been observed [95].

Homologous recombination can be defined as a process of genetic exchange between sequences that share homology. It is used by the cellular DNA repair machinery to repair a broken DNA molecule with a homologous sequence from a sister chromatid, homologous chromosome or a repetitive sequence for instance. The most widely accepted models for homology dependent DSB repair pathways include single strand annealing (SSA), synthesis dependent single strand annealing (SDSA), the double-strand break repair (DSBR) model, and the break induced replication (BIR) model.

The SSA model, initially proposed by Lin *et al.* [96], describes a frequent repair event between repetitive sequences with a recombination product containing a deletion. Single stranded resection products of two direct repeats interact with each other so that one copy of the repeats and the intervening sequence gets lost (Figure 6). It is thus a non-conservative process [89, 92, 93].

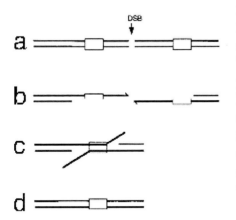

Figure 6: SSA model for DSB repair. Ends of a DSB between repeated sequences (boxes in a) are processed by extensive 5' to 3' resection to produce substantial regions of homology which can extent to 400 nt are exposed on ssDNA 3' ends [93] (b). Strand annealing of the repeats flips out the regions in-between (c) which are removed by an endonuclease [92]. The gaps are then filled and ligated (d). Modified from Paques and Haber [97].

One of the most widely accepted models for repair of DSBs via homologous recombination is the DSBR model developed from Szostak *et al.* [98]. As shown in Figure 7, a protruding 3' single-stranded DNA end invades a homologous sequence and displaces one strand of the original sequence. This displacement loop (D-loop) is enlarged by DNA synthesis initiated from the invasive end until the second 3' DNA end can hybridise to the D-loop and act as a primer for a second round of DNA synthesis. Through progression of this reaction two Holliday junctions are formed that can be resolved in two ways resulting in either crossover, or non-crossover. Therefore, it is a semi-conservative process associated with crossover events if the double Holliday junctions are resolved at different cleavage sites, although these events are suppressed in mitotic cells [92]. The model has proven to be suitable to explain typical products of gene targeting via homologous recombination [99].

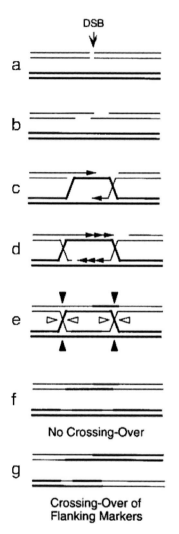

a

b

c

d

e

f

No Crossing-Over

g

Crossing-Over of
Flanking Markers

Figure 7: Double-strand break repair (DSBR) model according to Szostak *et al.* [98].

After introduction of a DSB (a) 3' single-stranded DNA (ssDNA) overhangs are produced via 5' to 3' exonucleolytic resection of DNA ends (b). Strand invasion of a protruding 3' single-stranded DNA end into a homologous sequence and formation of a D-loop (c) is mainly mediated by RAD51 that, among other things, forms a nucleoprotein filament on ssDNA and catalyses the search for homologous sequences. DNA synthesis, catalysed by a DNA polymerase, is initiated from the invasive 3' end and proceeded through the other end of the DSB. The D-loop is enlarged concurrently and a Holliday Junction is left at the site of invasion and may branch migrate (d). The D-loop serves as a template for DNA synthesis from the second 3' end and a second Holliday Junction forms after hybridisation of the first invasive 3' end on the other side of the DSB (e). The double Holliday Junction structure may be resolved by cutting inner (crossed) strands (open arrowheads) on both sides what results in non-crossover (f) or crossover (g) following inner or outer strand (filled arrowheads) cleavage on either side. Reviewed by Helleday [92], figure modified from Paques and Haber [97].

DSB repair based on SDSA is similar to the DSBR model and also capable of repairing short tracks, but newly synthesised strands are displaced from the template prior formation of a second Holliday junction. The two 3' ends of the lesion hybridise, gaps get filled and single strand breaks are ligated. This turns out in a conservative recombination event without crossover in most cases. Two simple forms of SDSA are outlined in Figure 8.

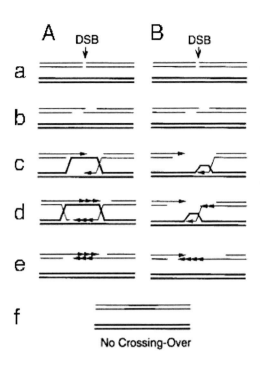

Figure 8: SDSA models for DSB repair.

DNA ends of a DSB (a) are processed to form ssDNA 3' overhangs (b) like in the DSBR model. After strand invasion (c) DNA synthesis is initiated at invasive 3' ends (d). Newly synthesised strands are displaced from the template and allowed to hybridise to each other (e). This is permitted by either actively splitting of the replication structure in the simple SDSA model (A), or if the D-loop remains small in the bubble migration model (B). Single stranded stretches are filled and linked resulting in non-crossover recombination (f). Modified from Paques and Haber [97].

The BIR model plays an important role in the repair of telomeric DSBs where only one end of the break may be available. It was originally formulated as 'break-and-replicate' model by Voelkel-Meiman and Roeder [100] and includes the formation of a true replication fork that might proceed to the end of the homologous chromosome allowing the acquisition of telomere sequences. Figure 9 shows the central reaction of this model, which incorporates many features of the DSBR model. Including the prerequisite of a DSB and end processing, the first steps are identical to the DSBR model with the exception that one end of a DSB is sufficient. Synthesis of a single-stranded product is initiated from an invasive 3' DNA end in a D-loop of limited size. Coordinated lagging strand synthesis converts the single-stranded product into double-stranded DNA. Synthesis can either proceed to the end of the chromosome, or be converted into a gap repair if the second end of the DSB becomes involved. The outcome of the reaction is conservative DNA-synthesis where the recipient chromosome has two newly synthesised strands [93, 101].

Figure 9: Central reaction of the BIR model for DSB repair.
Major feature of the BIR model is that only one end of the DSB invades the homologous sequence and initiates both leading and lagging strand synthesis within a D-loop. The size of the D-loop is limited due to branch migration following the direction of leading strand synthesis. Coordinated lagging strand synthesis converts the single stranded product of leading strand synthesis into a double strand. Reviewed by Flores-Rozas and Kolodner [101] and Pfeiffer *et al.* [93]; figure modified from Paques and Haber [97].

Although NHEJ is frequently used for DSB repair in mammalian cells, the prevalence for DSB repair by illegitimate recombination or homologous recombination pathways depends on the cell cycle phase, as reviewed earlier [91, 93, 102]. Owing to a decrease in Ku-proteins, which participate in NHEJ, and the availability of sister chromatids after replication homologous recombination is most efficient in late S and G2 phase of the cell cycle. SSA and NHEJ are expected to be used mainly during G1 phase, but are not restricted to this phase.

30% to 50% of all DSB in CHO cells are repaired by homologous recombination pathways [95, 103].

4 Results and discussion

This work describes the development of a CHO-K1 based expression system for targeted integration of genes encoding proteins of interest, herein called CEMAX system (cellular meganuclease assisted expression system). Based on modified CEMAX host cells any gene of interest can be integrated at a predefined genomic site by means of DNA double-strand break induced homologous recombination. The following steps towards implementation of the expression system were performed in this work:

1. Development of a concept of the recombination system, establishment of a cloning strategy and cloning of tag and replacement vectors

2. Generation of modified host cells (CEMAX host cell) by screening for a genomic locus that allows *in vivo* recombination and high expression levels

3. Site-specific integration of the gene for green fluorescent protein as a proof of concept

4. Generation and characterisation of producer cells (secreted model proteins)

Additionally, a problem that evolved during this work had to be solved. This aspect was the reestablishment of Celonic's serum-free cloning and expression technology after dilution cloning became impossible through a change in medium formulation. Therefore commercial available culture media were assessed for their potential to promote single cell growth.

4.1 Development of a concept and cloning of the vector system

Targeted integration means the targeting of an expression cassette for a gene of interest to a specific genomic locus where it is incorporated into the genomic DNA. In the context of protein expression this locus should allow high and stable levels of transcription. To target an expression cassette to a specific locus, the locus has to be modified with recognition sites for an enzyme catalysing the integration or at least inducing cellular mechanisms. This section describes the development of the vector system that was a prerequisite for enabling targeted integration in cultured CHO cells. In addition to a transient source of homing endonuclease activity the system comprises a tag vector and a replacement vector. During the development of CEMAX host cells the tag vector allowed placing a target site by modifying an endogenous locus and screening for high expression rates. The replacement vector, which was used in order to generate producer cell lines by targeted integration, facilitates replacement of the reporter cassette at the target site for the gene of interest. Once a frozen stock of CEMAX host cells containing the artificial target site at an appropriate locus is established, it could be used for reproducible, targeted integration of various genes.

This work was focused on gene targeting via DNA double-strand break induced homologous recombination. DNA double-strand breaks in the target site, which are introduced by the rare cutting homing endonuclease I-SceI, stimulate the cellular DNA repair machinery. One option to repair breaks in DNA is homologous recombination with an intact copy of the damaged region. This repair matrix can either be the second allele in case of a natural locus, or extrachromosomal DNA provided in form of a replacement vector for an artificial target site. The replacement vector is introduced into the cell via means of transfection and contains homologous sequences to the target site. These homologous sequences flank an expression cassette for the gene that should be integrated. As long as the artificial target site is introduced as a single copy, the replacement vector is the only available repair matrix, which allows targeted integration of the expression cassette by homologous recombination. An I-SceI induced recombination system was chosen since gene targeting via homologous recombination is a scar less and irreversible process and thus the potential for genome engineering is higher than that of recombinases. Recombinases have target site requirements on both molecules involved in recombination. These sites remain after recombination and make the reaction on one hand reversible [77]. On the other hand, these sites have to be placed flanking the replacement cassette and may interfere with transcription when positioned between a promoter and the gene of interest for instance. Regarding recombination frequency both systems should perform similar. As reported for Cre recombinase, frequency is about $5 \cdot 10^{-6}$ events per treated cell [104] and comparable to 10^{-5} to 10^{-6} events per treated cell with homing endonuclease induced gene targeting [61, 73, 105].

A CEMAX host cell had to be modified genetically with the tag vector to create an artificial target site at a locus, which allows high and stable rates of transcription. The tag vector should therefore have the following essential features:

- A reporter gene for screening of highest expression rates to allow identification of suitable locus for site-specific integration

- Sites for introduction of double-strand breaks by I-SceI

- Regions of homology to the replacement vector

- Elements that allow selection of homologous recombination

The replacement vector, which provides the repair matrix and the gene of interest, should have the following features:

- Regions of homology to the artificial target site

- Selection markers that get activated through recombination

- A cloning site for the gene of interest

- Ideally absent promoter activity of the expression cassette in the context of the replacement vector

Generation of producer cells requires a target site in the genome of the host cell. In a first step this recombinant host cell, which contains the artificial target site, had to be developed. Based on frozen aliquots of these cells a cell line producing the protein of interest could be generated in a reproducible manner. Growth characteristics and production capacities of producer cells should be roughly predictable.

4.1.1 Blueprint for the vector system and recombination scheme

Central elements of the CEMAX system are plasmid vectors that allow homologous recombination at a selected locus in the genome of the host cell.

These vectors and the scheme for targeted integration via double-strand break induced homologous recombination are illustrated in Figure 10.

The tag vector was designed to introduce a genomic tag into the host cell (section A in Figure 10). This tag includes recognition sites for the homing endonuclease I-SceI. Between these recognition sites lies an exchangeable cassette including a reporter gene coding for human secreted alkaline phosphatase (hSEAP) and a hygro gene for selection of stably transfected cells that produce the reporter gene at high expression levels. Regions that are homologous to the replacement vector flank these elements. On the 5'-end this homology comprises the intron of the CMV promoter and on the 3'-end the neomycine phosphotransferase gene that lacks a promoter and the start codon (Δneo) and is thus inactive in the context of the tag vector.

The replacement vector (section B in Figure 10) was used to introduce the gene of interest into the CEMAX host cell in order to generate a stable producer cell line by targeted integration. The gene of interest in the replacement vector is flanked by homologous sequences to the genomic tag to allow targeted insertion via homologous recombination. On the 5'-end this homology comprises the first exon and intron of the CMV promoter. The 3'-region of homology is a truncated expression cassette of the neomycine phosphotransferase gene (neoΔ). This version includes a promoter and a start codon but lacks the terminal 161 base pairs (bp) of the resistance gene. This neoΔ should not confer resistance to G418 since a deletion of the terminal 74 bp (3´end) results in a nonfunctional protein [106] and a missing polyadenylation signal would interfere with mRNA processing.

Section C in Figure 10 illustrates the recombined genomic locus of a CEMAX producer cell. The initial reporter and selection cassette is exchanged for the gene of interest that is expressed under control of the CMV promoter, which was retained during recombination. There is a marker for selection of homologous recombination events on each end of the exchangeable expression cassette. This is a Zeocin resistance gene (zeo) linked to the GOI via an IRES on the 5'-end and the restored neo expression cassette on the 3'-end respectively. The activation mechanism of the neo is analogous to a plug and socket strategy described earlier for the hypoxanthine phosphoribosyltransferase selection system [59]. The Zeocin resistance gene was introduced in the vector design after first gene targeting experiments have shown that it was not sufficient to solely select with the neo marker (see section 4.2.1.3). This additional marker and the GOI are only expressed under control of the CMV promoter after homologous recombination since there is no promoter activity in the context of the exchange vector.

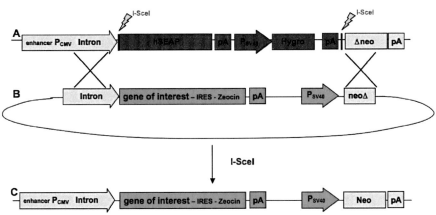

Figure 10: Recombination scheme of the CEMAX system.
A: tag vector integrated in the host cell genome, B: replacement vector containing the gene of interest, C: recombined locus in the genome of the host cell. Crosses indicate the homologies between the tag vector and the replacement vector. Elements that remain after recombination are coloured in grey. The reporter and selection cassette in the tag vector (red elements) is used for selection of high producer host cells and is lost upon recombination. The replacement cassette containing the gene of interest (green elements) is integrated by homologous recombination catalysed by the cellular DNA repair machinery. This process is induced by DSBs introduced by I-SceI. P_{CMV}: major immediate early promoter of hCMV; P_{SV40}: SV40 early promoter; pA: polyadenylation signal; Δneo: neo gene lacking a promoter and initiator codon; neoΔ: neo expression cassette lacking terminal codons and polyadenylation signal. Adapted from Greulich et al. [107, 108].

The total length of homology, which extends over 1598 bp consisting of 5´homology of 970 bp in the promoter region and 3´homology of 628 bp in the neo region, is approximately 1000 bp shorter than homologies published earlier [61, 73]. Possible reduction of recombination frequency due to short homologies might be compensated by the use of two I-SceI cleavage sites. Less processing at the double-strand break will be necessary since both DNA ends are exposed near the regions of homology after cleavage at two sites. In Addition two sites will help to overcome limitations in the length of exchangeable cassettes [90].

4.1.2 Cloning of the CEMAX vector system

With regard to the requirements of the vector system described above, the cloning strategy was developed based on proprietary and commercial available plasmid vectors. Synthetic DNA fragments and oligonucleotides were used to introduce sequences that were not incorporated in available plasmids or to amplify DNA fragments by polymerase chain reaction (PCR). This section provides a brief overview of the cloning strategy. Figure 11 shows the cloning steps to obtain the tag vector CV050. In the first step the I-SceI recognition site was introduced into the expression vector CV001. Hybridising synthetic oligonucleotides to a HindIII compatible double-stranded DNA fragment and inserting it into the HindIII site of CV001 enabled this step. Further modifications were the introduction of the hygro gene, the insertion of the second I-SceI recognition site linked to the promoter less Δneo and cloning the reporter gene hSEAP. Two additional cloning steps were necessary due to a frameshift mutation in the hygro gene, which was unfortunately introduced via a mutated PCR primer. First, the mutated gene was exchanged for a new PCR product lacking the mutation. In the second step the reporter gene hSEAP, which was deleted through the first step, was inserted again. The tag vector CV050 was then used for the generation of a first cell line generation.

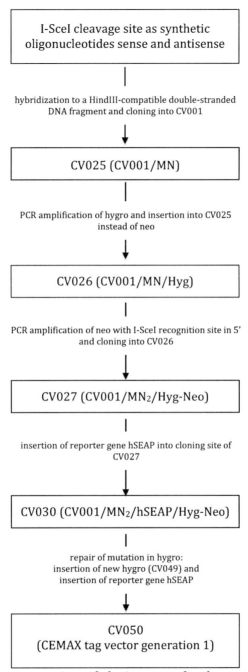

Figure 11: Schematic overview of cloning steps for the generation of the tag vector CV050.

A frameshift mutation in the hygro marker was introduced by a mutated PCR primer was removed in two additional cloning steps prior use of the vector.

Vector elements for the replacement vector, which was used to transfer the gene of interest into the CEMAX cell, were combined as described in Figure 12.

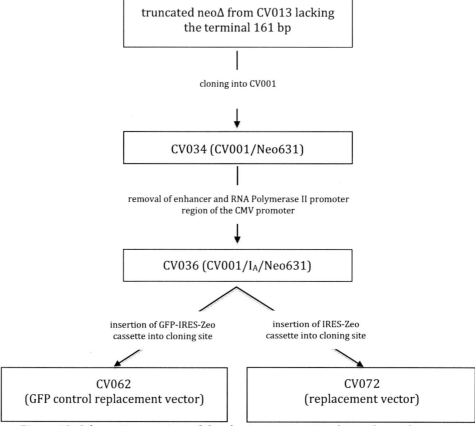

Figure 12: Schematic overview of the cloning strategy to obtain the replacement vectors.
GFP was used as a reporter in the control replacement vector.

Briefly, a neoΔ lacking the terminal 161 bp was inserted into CV001 under control of the Simian Virus 40 (SV40) early promoter. The second cloning step comprised the removal of CMV promoter elements. The remaining elements should provide enough homology for recombination within the CEMAX host cell and the deletion, which includes the transcription initiation site, should prevent promoter activity. This was achieved by deleting the enhancer and the RNA polymerase II promoter region but leaving the first exon and intron as homologous sequences to the CMV expression cassette in the tag vector. A control replacement vector was then constructed by inserting the gene for the green fluorescent protein (GFP), an internal ribosome entry site (IRES) derived from foot and mouth disease virus (FMDV) and the zeo gene into the cloning site. A replacement vector for cloning a GOI (CV072) was constructed similarly but harbours a cloning site instead of the GFP gene.

Plasmid maps of the tag vector CV050 and the replacement vector CV072 are depicted in Figure 13 and Figure 14, respectively. The PsiI restriction site within the replacement cassette of CV072 could be used for insertion of an additional expression cassette. This expression cassette should not contain a CMV promoter, which could result in competitive recombination events due to homologies with the tag vector. A second expression cassette is necessary for expression of antibodies or other proteins containing two different peptide chains.

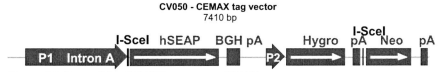

Figure 13: Map of the CEMAX tag vector CV050.
The map shows the linear form of the vector after treatment with SspI which was used for transfection. With P1: hCMV major immediate early promoter with enhancer and Intron A; I-SceI: recognition site of I-SceI; hSEAP: human secreted alkaline phosphatase; BGH pA: polyadenylation signal of bovine growth hormone gene; P2: SV40 early promoter; Hygro: hygromycin phosphotransferase gene; pA: SV40 early polyadenylation signal; Neo: Δneo.

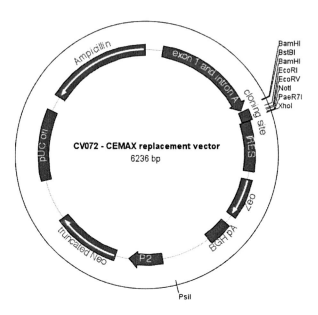

Figure 14: Map of the CEMAX replacement vector CV072.
A gene of interest can be inserted into the cloning site of CV072. If a second expression cassette is necessary it can be introduced into the PsiI site within the replacement cassette. With IRES: internal ribosome entry site of FMDV; Zeo: Zeocin resistance gene; BGH pA: polyadenylation signal of bovine growth hormone gene; P2: SV40 early promoter; truncated neo: neoΔ; pUCori and Ampicillin: origin of replication derived of cloning vector from University of California and ampicillin resistance gene for plasmid propagation in *E.coli*.

4.2 Cell line development for generation of CEMAX host cells

This section describes the generation of modified host cells that allow targeted integration of expression units. These host cells harbour a reporter cassette at an endogenous hot spot of expression, which was introduced through the tag vector. Afterwards, the reporter cassette was replaced by a newly introduced expression unit, which was delivered via the replacement vector upon targeted integration (section 4.3). Host cell modification was done with the tag vector, which provided recognition sites for the homing endonuclease I-SceI, sequences for homologous recombination with the replacement vector, and elements for selection of stable high producer clones (Figure 13). Clones having this tag stably integrated at a transcriptional highly active site were referred to as CEMAX host cells.

The first part of this section deals with clone screening of cell lines based on 14-CHO-S cells transfected with the tag vector CV050 and a first gene targeting experiment (4.2.1). The following section deals with the development of a second series of CEMAX host cells using the modified tag vector CV063 for screening of additional host cell candidates with a slightly modified tag vector design (4.2.2). Theoretical considerations regarding the effect of a time point when a cloning experiment is initiated on success-probabilities to pick particular clones are presented in appendix 8.1. Taking into account these theoretical considerations, the cloning experiments were initiated on one hand at the day after transfection while applying selection pressure immediately. This strategy combined limited dilution cloning with selection of stable cells in a mini pool approach. In addition bulk pools of stable transfectants were selected in suspension for the first generation of host cells. Cloning experiments were then performed with stable cells from the bulk pool. In the case of cell line generation 2 only mini pools were selected after this strategy was applied successful for detection of outperformer clones in the first experiments. A higher throughput method for selection of a third and fourth generation of host cell candidates is described in appendix 8.2.

The cell lines were designated according to the host cell line used for transfection and the plasmid introduced into the cell. 14-CHO-S/CV050, for example, represents a bulk pool of 14-CHO-S cells transfected with the plasmid vector CV050. Clones were named and numbered according to the cloning experiment followed by an individual clone number within this experiment. Accordingly clone 1 from cloning experiment LD02 is termed clone 02-001. Clones referred to as outperformer clones were characterised by a high activity of hSEAP in cell culture supernatant of 96-wells that was above the quantification limit under test conditions. The screening program was designed to discover such clones. In addition to high productivity it was essential that the high expression level was conserved after gene replacement at the target locus. This was analysed in gene targeting experiments with genes encoding model proteins such as GFP and secreted proteins.

4.2.1 CEMAX host cell line generation 1

The first series of CEMAX host cell clones was based on the tag vector CV050 and the host cell line 14-CHO-S growing in BD CHO medium. The development strategy (Figure 15) included generation of stable clones by the mini pool approach and by limited dilution cloning of stable cells selected as bulk pools. These clones were then submitted to a second cloning experiment to ensure clonality and were tested for gene targeting and functional integration with the GFP replacement vector. Selected host cell clones were characterised for productivity of hSEAP and analysed in Southern blots to determine the integration pattern of the tag vector.

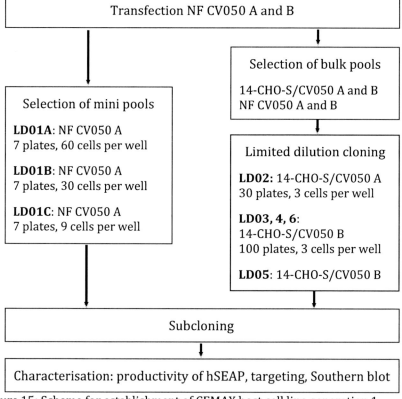

Figure 15: Scheme for establishment of CEMAX host cell line generation 1.
Mini pools and bulk pools were selected based on two transfections with different quantities of the tag vector CV050. 1 pmol plasmid DNA was used in experiment NF CV050 A and 0.02 pmol in experiment B.

4.2.1.1 Clone screening

Stably transfected high producer clones are rare in a pool of weak producers due to the random nature of transgene integration. These high producers were in the focus of this clone screening approach to obtain clones having the reporter cassette integrated at an endogenous hot spot of expression.

To achieve this, 14-CHO-S cells were transfected with two different amounts of the tag vector CV050 and submitted to mini pool and bulk pool selection (Figure 15). In experiment NF CV050 A cells were transfected with 1 pmol CV050, whereas 0.02 pmol plasmid DNA were used for NF CV050 B. This reduced DNA amount used for transfection should theoretically lead to integration events with lower transgene copy numbers. Low copy numbers were essential since several I-SceI cleavage sites in the genome might cause genome rearrangements [89]. Several copies of the tag vector integrated as tandem arrays would be reduced to a single copy upon cleavage with I-SceI. Cells from transfection experiment A were selected as both bulk pools and mini pools, whereas cells from experiment B were selected as bulk pools prior cloning via limited dilution. Selected mini pools were then subcloned to attain clonal cell lines before the clones were characterised.

The bulk pool selection of experiment NF CV050 A and B is depicted in Figure 16. Up to day 3 both cultures exhibited an identical recovery from transfection. The effect of selection compound hygromycin B became apparent between day 3 and 7 followed by considerable differences in the course of selection owing to different frequencies of stable cells within the pools. In experiment A viability dropped to 60% accompanied with a reduction in growth rate and recovery on day 10 by outgrowth of resistant cells. Time for recovery took much longer in experiment B. A mass of untransfected and unstable cells was eliminated in a 15 days lasting decline in viability and viable cell numbers until outgrowth of resistant cells was detectable from day 18 on. Subsequently viability increased by the outgrowth of viable cells. The increase in viability on day 15 was most probably due to an inaccuracy in counting, which was consistent with low cloning efficiency obtained in cloning experiment LD03 (see Table 2).

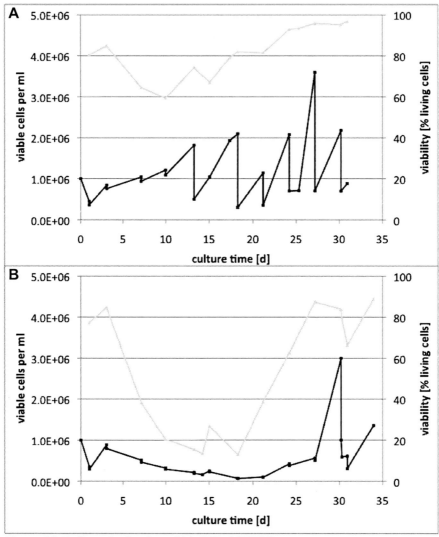

Figure 16: Bulk pool selection of 14-CHO-S/CV050.

Two data in viable cell density (black line) at one point in time, connected through a vertical line, indicate dilution of cells. The light grey line shows progression of viability. A: Bulk selection of 14-CHO-S/CV050 NF CV050 A (1 pmol CV050). The cells were selected in 6-well scale between day 0 and 13, expanded in T-flasks until day 21 and then cultivated in a spinner flask. B: Bulk selection of 14-CHO-S/CV050 NF CV050 B (0.02 pmol plasmid DNA). This experiment was selected in 6-well scale until day 30 and then in a T-flask. Cloning experiment LD01 was initiated on day 1 and LD02 on day 13 from experiment NF CV050 A. LD03, LD04, LD05 and LD06 were performed with cells from experiment NF CV050 B on day 15, 35 and 37 respectively.

The selection of mini pools as a combination of limited dilution cloning and selection of stable cells was performed in parallel to the bulk pool selection with cells from experiment A (Table 1). Stable cells occurred at a frequency of $1.9*10^{-2}$ per transfected cell as calculated from this mini pool approach. The other cloning experiments were performed with preselected cells of experiment A on day 13 (LD02) and of experiment B on day 15, twice on day 35, and on day 37 (LD03 to LD06). Table 2 summarises the limited dilution cloning experiments and results of these. The cloning efficiency varied between 0.2% and 2.4% in these experiments, which could be an effect of the condition of the parental culture.

The mini pool approach had several benefits compared to the limited dilution approach. It was not necessary to select a bulk pool and the yield of clones per plate was higher than with the limited dilution approach. In addition to that, clones were at least two weeks earlier available.

Table 1: Overview of screening experiments by the mini pool approach.
LD01 represents the mini pool approach and was inoculated 24 h after transfection of 1 pmol CV050 (NF CV050 A) for selection in wells of a microplate. Although showing lower expression levels clones 01B001 and 01B263 were used for comparison with outperformer clones in gene targeting experiments. The frequency of stable clones was calculated to 1.9±0.5% from cloning experiment LD01A, B and C.

Cloning experiment	Number of microplates	Cells per well inoculum	Number of clones	Outperformer clones
LD01A	7	60	494	01A438
LD01B	7	30	414	01B250
LD01C	7	9	139	01C090

Table 2: Overview of screening experiments by limited dilution.

All experiments were performed with preselected bulk pools of stable cells. LD02 was performed from NF CV050 A. LD03 to LD06 were performed after transfection with reduced DNA amount (NF CV050 B). Although showing lower expression levels clones 03-002, 03-022, and 04-106 were used for comparison with outperformer clones in gene targeting experiments. The cloning efficiency varied between 0.2% and 2.4%.

Cloning experiment	Number of microplates	Cells per well inoculum	Number of clones	Outperformer clones
LD02	30	3	204	02-117, -124 and -202
LD03	30	3	47	No outperformer. Low level of expression.
LD04	50	3	131	
LD05	20	10	435	
LD06	20	3	9	

In total 1873 clones were analysed for the activity of the reporter protein hSEAP in the supernatant of 96-wells after growth to 20% to 100% confluence. The expression level of clones derived from bulk pools transfected with 0.02 pmol DNA showed considerably less expression of the reporter gene compared to outperformer clones transfected with 1 pmol CV050. This might be due to lower copy numbers, incomplete integration of plasmid DNA and overgrowth of clones with low productivity during the increased period of bulk pool selection. Limited dilution cloning seems to be not applicable for approaches like this, which might require higher throughput methods like cell sorting for instance. Such a screening program was used for isolation of high producers during development of a third and fourth host cell generation and is described in appendix 8.2.

Figure 17 exemplifies the result of the analysis of mini pools in cloning experiment LD01C. In this experiment the outperformer clone 01C090 was chosen for further experiments.

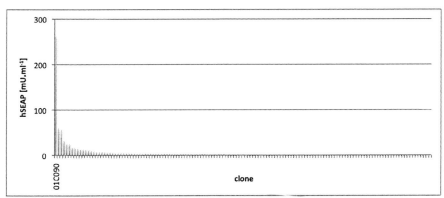

Figure 17: SEAP activity in 96-well supernatant of mini pool selection LD01C. Outperformer clone 01C090 was the best clone among 139 clones from this experiment.

Subcloning experiments were performed with shortlisted clones as specified in Table 3 and clones with the highest productivities in the independent experiments were selected for assessment in gene targeting experiments. Subclones of clone 01A438 showed similar high level of SEAP activity in supernatant as well as subclones of 01C090, which might indicate an already clonal parental culture. Clone 08-018 was the only outperformer subclone of 01B250. Subclones of 01B001 and 01B263 showed levels of SEAP activity that were in the range of the mini pool used for subcloning indicating that there was no high producer in these parental cultures.

Table 3: Overview of subcloning experiments.
Plates were analysed for growth of single cell derived clones to ensure clonality.

Cloning experiment	Mini pool	Selected clones
LD07	01A438	07-022
LD08	01B250	08-018
LD09	01B263	09-002
LD10	01B001	10-090
LD26	01C090	26-006, 26-007, 26-008 and 26-010

In addition to outperformer clones, five other clones derived from both transfections were chosen for further cell expansion and characterisation. 01B001, 01B263, 03-002, 03-022, and 04-106 exhibited significantly lower productivity than outperformers. For example these clones produced between 16 and 100 mU·ml^{-1} hSEAP in 96-well scale. Outperformer clones produced more than 250 mU·ml^{-1} in the 96-well supernatant. Cell specific productivity of initially selected clones was determined subsequently as described below. After freezing of a small primary seed bank (PSB) shortlisted clones were chosen for comparison of low and high producer host cells regarding GFP expression after targeted integration.

4.2.1.2 Productivity of secreted alkaline phosphatase

The accumulation of hSEAP in supernatant during cell cloning was used as a rough indicator to estimate productivity of a huge number of clones. The rate of production in micro unit per cell and day ($\mu U \cdot c^{-1} \cdot d^{-1}$) was then determined either during cultivation or in 6-well assays. Results were used as a selection criterion among initially selected mini pools and clones. This more precise analysis of production rate for the best clones achieved is shown in Figure 18.

The highest productivity among clones of generation 1 exhibited mini pool 01C090 and clone 26-010, which is a subclone of 01C090. Productivity of 01C090 and 26-010 was roughly the same, which is consistent with the fact that 01C090 subclones produced similar 96-well hSEAP concentrations in the subcloning experiment. Clone 07-022 lost 50% of its productivity during cultivation for 53 generations in both nonselective medium and selective medium (detailed data not shown). This indicated instability of production that might be caused by, for example, genetic instability, gene silencing by DNA methylation, or impaired posttranscriptional processes [109] that manifested during prolonged cultivation. Reasons for differences in productivity among clones may be manifold and include transgene copy numbers and position effect at the integration site [110]. But, at the end of the day, productivity of a secreted biopharmaceutical product candidate or another relevant protein after targeted integration counts most.

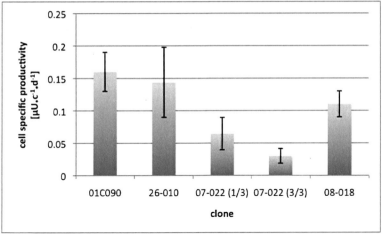

Figure 18: Cell specific productivity of selected CEMAX host cell clones. Productivity of clone 07-022 dropped during prolonged cultivation to approximately 50% of the initial productivity. Mean values were calculated from data of the first (1/3) and last third (3/3) of the prolonged cultivation of 07-022. Error bars indicate variations in productivity during cultivation.

4.2.1.3 Survey of site-specific and functional integration

After screening of host cells for high specific productivity, shortlisted clones were analysed for their competence to allow replacement of the reporter cassette for an expression unit containing a GOI. Productivity should be conserved at high level after gene targeting. Cotransfection experiments with the GFP control replacement vector and the I-SceI expression plasmid were performed to survey this. Therefore, $2 \cdot 10^6$ cotransfected cells were either subjected to G418 selection or G418 and Zeocin double selection in serum-containing medium on 150 cm^2 tissue culture dishes. The development and optimisation of a gene targeting and selection protocol is described in detail in section 4.3.1.

Several clones identified as high producers in the cloning experiment described above, were not usable for targeted integration and were excluded from further experiments. These were clones 02-117, 02-124 and 02-202. The same was the case with low producer clones 04-106, 09-002, 10-090 and manifested itself in failure to select G418 and Zeocin resistant cells after cotransfection. Possible reasons for that could have been high transgene copy numbers and several integration sites that caused genome rearrangements. This lethal effect was described earlier for DSBs introduced on different chromosomes [89]. This could be the case in particular for high producer clones. Lower accessibility of the recombination machinery to the genomic tag due to heterochromatin environment [57, 62] as well as incomplete integration of the tag vector and thus failure to activate selection markers could be the reason especially for low producer clones.

The high producer clones 07-022, 08-018 and 01C090 and subclones of 01C090 showed high expression levels of GFP expression after targeted integration of the control replacement vector. Figure 19 illustrates examples of GFP producer cells generated by targeted integration via DNA double strand break induced homologous recombination. These clones occurred at a frequency in the range of 10^{-5} to 10^{-6} per transfected cell. The gene targeting experiment with host cell 07-022 (Figure 19, right picture) showed that it was not sufficient to select targeted clones with G418 alone. Although GFP positive cells were selected, these cells were contaminated with non-producer cells that were resistant to G418.

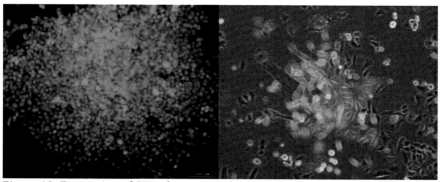

Figure 19: Expression of GFP after targeted integration in generation 1 host cells. CEMAX host cells 01C090 (left) after targeted integration of the GFP gene from the control replacement vector CV062 and double selection with G418 and Zeocin (UV light exposure with the GFP filter setting). Host cell clone 07-022 (right) was subjected to G418 selection (UV and VIS light with the GFP filter setting). The bars indicate 100 μm.

A control experiment was performed to analyse the effect of GFP negative G418 resistant cells. CEMAX host cell clone 01C090 was transfected with the I-SceI expression plasmid followed by G418 selection. The frequency of G418 resistant clones in this experiment was above $2.5 \cdot 10^{-4}$ events per transfected cell and significantly higher than the frequency of correctly recombined clones.

Reasons for resistant non-producers could have been competitive recombination events to homologous recombination. These include activation of the neo marker via end joining. End joining connects the neo gene with the CMV promoter and generation of an AUG initiator codon during end processing generates a functional neo expression cassette. Alternatively translation initiation from non-AUG initiator codon might allow expression of the Δneo [111, 112]. This could be caused by the tag vector design with a promoter on the one end and the Δneo marker on the other end of a DSB. With a similar vector design neo activation was observed although the Δneo had a 20 bp deletion on the 5' end of the gene as personally communicated by Dr. Cabaniols [113]. It has been reported that either small deletions, or insertions of up to 205 bp from distinct sequence elements were observed during end joining [95]. A second possibility for neo activation in gene targeting reactions might be a semi-illegitimate recombination reaction [72, 74, 99]. Homologous crossover on one end of the replacement cassette accompanied by a nonhomologous event on the other end would result in partial gene replacement at the target locus. In this scenario one end of the cassette would span recombined sequences while the other comprises original sequences. These events would confer resistance to G418 if the neo end contains sequences from the replacement vector while expression of the GOI is not detectable. Referred to as "one-sided" recombination this mechanism is consistent with other findings and has been found to predominate over homologous recombination events in mouse 3T3 cells [72] while it is contrary in mouse ES cells [74]. The issue of G418 resistant non-producers after cotransfection was addressed in two strategies. A Zeocin resistance marker was linked to the GOI in the replacement vector via an IRES. Cotransfected cells could thus be submitted to G418 and Zeocin double selection, which allowed selection of homologous recombination events on both sides of the exchangeable region. On the other hand the tag vector design was modified for development of a second series of CEMAX host cells to reduce the probability of activation of the Δneo through end-joining (see 4.2.2).

One predicted feature of the expression system was confirmed by subjecting low producer host cells to targeting reactions. As expected the expression level of GFP expression from these host cells was low compared to cells derived from the three high producer clones. Clones 03-002 and 03-022 were examples of low producer host cell clones that showed low levels of GFP expression after gene targeting. Although differences in GFP fluorescence between high and low producers were detectable by fluorescence microscopy, the comparison of high producers either requires standardised analysis using flow cytometry for example, or analysis of secreted proteins that can be quantified accurately.

Promising host cell clones were analysed in more detail after a successful gene targeting experiment with the GFP replacement vector. The integration of the tag vector was characterised by Southern blot analysis.

4.2.1.4 Analysis of the integration pattern of the tag vector

Southern blot analysis was used as a method for rough characterisation of the integration pattern of the tag vector. Since several copies of the tag vector containing I-SceI cleavage sites distributed through the genome of a cell might cause lethal genome rearrangements [89], a single and intact copy of the tag vector would be the best case scenario. A cooperation partner performed the analysis. Genomic DNA of clones and mini pools was treated with restriction endonuclease cutting sequences of the tag vector prior blotting and detection. Figure 20 shows restriction sites, sequences that hybridise with the probes and the minimum fragment sizes that will be obtained on the blot. Two probes, one at each end of the tag vector, were used to detect restriction fragments that contain vector sequences. The Southern blot analysis of CEMAX host cell generation 1 is shown in Figure 21.

No clones with an intact single copy integration of tag vector CV050 were detected. The integration pattern of clone 01C090 (sample block 1 in Figure 21), which was identical to the pattern of all subclones that were analysed (26-008 and 26-010 in sample block 2 and 3), was complex. Three fragments in addition to the control fragment were detected with the CMV probe from SacI treated DNA indicating three copies of the promoter region. Two fragments from SacI treated DNA were detected with the probe corresponding to the neo region of the tag vector. Thus, this clone might contain two complete integrations and additionally a fragment of the tag vector that includes the CMV promoter region but not the Δneo.

Clone 08-018 (sample block 5 in Figure 21) showed 2 bands in each restriction on the neo blot, but only one band from each restriction was detected with the CMV probe. This observation could be interpreted as a complete integration of the tag vector combined with additional sequences from the neo region of the tag vector.

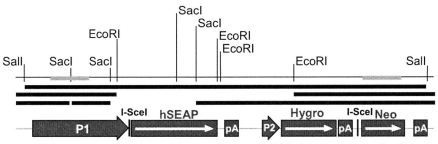

Figure 20: Restriction sites in the tag vector CV050 integrated as single copy integration.

Black bars indicate restriction fragments (from top to bottom: SalI, EcoRI, SacI) that were detected with the probes (grey bars). Minimum fragment sizes are with the probe corresponding to the CMV promoter: SalI: 7276 bp; EcoRI: 1883 bp; SacI: 1024 bp + 728 bp; and with the probe for detection of neo sequences: SalI: 7276 bp; EcoRI: 2274 bp; SacI: 4077 bp. P1: enhanced CMV promoter with intron A; P2: SV40 early promoter.

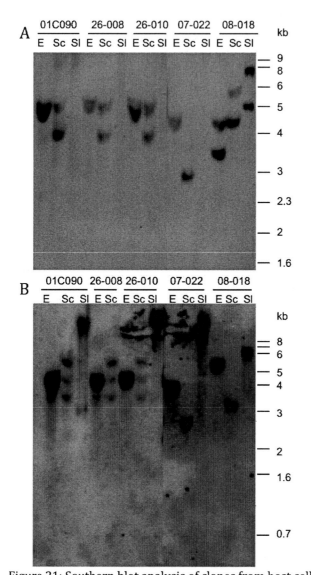

Figure 21: Southern blot analysis of clones from host cell line generation 1.
A: Fragments detected with a probe complementary to sequences of neo,
B: Detection with probe complementary to CMV sequences; E: EcoRI; Sc: SacI; Sl:
SalI. Minimum sizes after restriction with EcoRI are 2.3 kb and 1.8 kb for the neo
and the CMV probe respectively. For restriction with SacI this is 4.0 kb (neo
probe) and 1.0 kb plus an additional internal fragment of 728 bp, which was
barely detectable (CMV probe). Cleavage with SalI results theoretically in a
7.3 kb fragment detected with both probes. Fragment size from a DNA molecular
weight standard is indicated at the right side of the blots.

Interestingly in the context of successful targeting experiments with clone 07-022 (sample block 4 in Figure 21) was the result of the southern blot analysis of genomic DNA from this clone. The fragment detected with the neo probe in SacI treated DNA was 1 kb below the minimal size of 4.1 kb, which was calculated from the plasmid sequence. This could be either due to deletion of terminal sequences including approximately 600 bp of the neo gene, or an internal deletion of sequences at the 3'-end of the hSEAP gene. Deletion of neo sequences is unlikely since selection of G418 resistant cells after cotransfection was possible and would have been impaired by the deletion. Taking into account the loss of productivity after prolonged cultivation, which was observed for host cell clone 07-022 (Figure 18), genetic instability was the most probable reason for the result that was obtained in the Southern blot analysis.

The Southern blot analysis of genomic DNA treated with SalI had to be interpreted carefully for three reasons. First, the restriction sites are located at the ends of the linearised tag vector (Figure 20) and may be lost during integration [114, 115]. Secondly, cleavage activity of SalI that contains a CG in its recognition sequence is blocked by CpG methylation, and third concatemer integration would cause large fragments, if the SalI sites were lost. Thus, the occurrence of large fragments can be interpreted either as loss of terminal vector sequences during integration and formation of concatemers, methylation inactivation of the SalI site, or a combination of both. Since methylation of CpG islands is linked to gene inactivation as reviewed by Siegfried and Cedar [116], this might be a reason for the band pattern in the case of clone 07-022. This particular clone showed a smear above 10 kb in the Southern blots (Figure 21 B) indicating high molecular weight DNA and exhibited low and unstable productivity. Although this smear was of weak intensity on the neo blot, functional restriction sites for SalI seemed to be relatively rare around the integration site. Propagation of CpG methylation in the region of the integration site of the tag vector might be the reason for loss of productivity during prolonged cultivation.

Host cell clone 01C090 did not contain a single copy integration of the tag vector, but was expected to be clonal according to analysis of productivity and integration pattern. Subclones were identical among each other and provided no improvement compared to the parental clone. Thus 01C090 was used in subsequent gene targeting experiments owing to the advantage of approximately two months less time in culture compared to the subclones.

Although an intact single copy integration of the tag vector could not be detected, targeted integration of the gene encoding the GFP was possible with host cell clones 01C090, 07-022 and 08-018. These clones were used in further experiments to analyse the expression level of secreted proteins and to evaluate their applicability as expression hosts for production of biopharmaceutical proteins (see 4.3.2).

4.2.2 CEMAX host cell line generation 2

Cells that were resistant to G418 but did not produce the GOI were observed in the first attempts of site-specific integration with CEMAX host cell clones. This problem was addressed by two strategies. As described above, an additional selection marker was linked to the GOI in the replacement vector via an IRES. This section describes the second strategy to reduce the probability of neo activation through end joining by an improved tag vector design. The tag vector was modified by placing a transcriptional isolator between the 3' I-SceI site and the Δneo. Additional host cell candidates were selected by the mini pool approach based on 14-CHO-S cells transfected with this modified tag vector (CV063) and characterised as described for the first cell line generation.

4.2.2.1 Improvement of the tag vector

A transcriptional isolator was placed in the tag vector between the 3' I-SceI site and the Δneo in order to reduce the probability of transcriptional activation of the selection marker after I-SceI cleavage. To achieve this, the downstream I-SceI site was placed between the hygro gene and its polyadenylation signal, which should serve as a transcriptional isolator. Termination codons in all three reading frames were additionally inserted directly upstream of the Δneo. As for the first series of CEMAX host cells, the Δneo did not contain its native initiator codon. Activation of neo via end joining thus has to include steps as the removal of the transcriptional isolator during processing and generation of an initiator codon downstream the termination codons. This region was introduced into a precursor of the tag vector CV050 through a synthetic DNA fragment (Figure 22). To reduce gene synthesis effort the synthetic fragment was cloned into an intermediate vector and thereby linked to sequences of the hygro gene that were not included in the synthetic fragment. As before, hSEAP was chosen as the reporter gene in the tag vector CV063 for cell line generation 2.

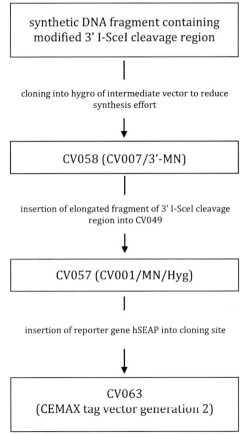

Figure 22: Cloning of the tag vector CV063.
The three-step approach was based on a synthetic DNA fragment containing the modified 3' cleavage region.

4.2.2.2 Clone Screening

As for the first series of host cell clones stable high producers that allow gene replacement at the target locus were in the focus of this screening approach. The selection of this second series of host cell clones was performed by the mini pool approach from three individual transfections of 14-CHO-S cells with the modified tag vector CV063 as outlined in Figure 23. It was omitted to select bulk pools after feasibility of the mini pool strategy was shown in cloning experiment LD01.

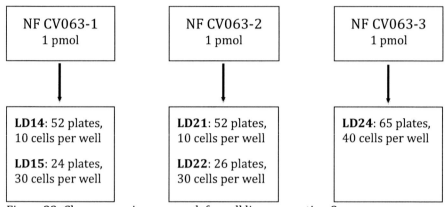

Figure 23: Clone screening approach for cell line generation 2.
Stable mini pools from three independent transfections were selected via limited dilution cloning 24 h after transfection.

The result of this clone screening program is summarised in Table 4 with the highest outcome of high producers obtained in experiment LD24. The result of cloning experiment LD24 is illustrated in Figure 24.

Table 4: Cloning experiments to obtain host cell line generation 2.
Clones with medium high level of SEAP activity in supernatant were recovered from cloning experiments LD14 and LD15. Higher expression levels were achieved in experiment LD24. Clones with low expression level were not investigated in additional experiments.

Cloning experiment	Number of clones	Further investigated clones
LD14	77	14-019
LD15	279	15-013, 15-152, 15-162
LD 21	61	low expression level
LD 22	345	low expression level
LD 24	3456	24-508, 24-1724, 24-1797, 24-2995, 24-3043

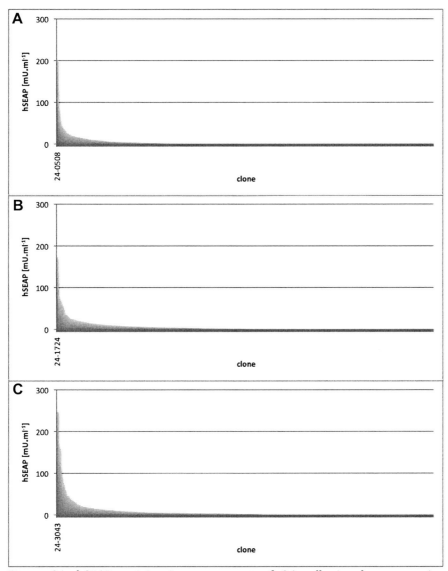

Figure 24: hSEAP activity in supernatant of 96-wells in clone screening experiment LD24.

In total samples of 3456 clones with minimum confluence of 40% were analysed in screening round A to C. A: clone 24-0001 to 24-1169, B: clone 24-1170 to 24-2744, C: clone 24-2745 to 24-3456.

In cloning experiment LD24 250000 transfected cells were plated and grew up to 3456 mini pools. This corresponded to a frequency of $1.4 \cdot 10^{-2}$ stable cells per cell after transfection. These mini pools were analysed for SEAP activity in 96-well plate supernatant in 3 screening rounds as shown in Figure 24. With three outperformer clones (24-508, 24-1724 and 24-3043) and two clones with approximately 30% lower expression level this screening approach was used successful for detection of high producer clones.

762 clones were analysed in addition to clones from LD 24, which made a total of 4218 mini pools from 5 cloning experiments. Variations in yield of mini pools were most probably based on differences in cloning efficiencies and deviations between individual transfection experiments.

Cell specific productivity was assessed during cell expansion. Afterwards, subcloning experiments of clones with the highest productivities were made to generate single cell derived cell lines. These experiments, which were done in parallel to gene targeting experiments, are summarised in Table 5.

Table 5: Overview of subcloning experiments from cell line generation 2.
Mini pools and subclones of 24-0508 and 15-013 were tested functional in GFP targeting experiments. *1: Discontinued after clone analysis since mini pools were either unstable or non-functional in gene targeting experiments. *2: It was focused on single cell derived subclones from LD31 due to higher cloning efficiency. Additional clones of LD31 were expanded and characterised in Southern blot experiments.

Cloning experiment	Mini pool	Selected clones
LD25	24-0508	25-003, 25-004, 25-006 and 25-007
LD27	24-1724	No clones*1
LD28	24-1797	No clones*1
LD29	24-3043	No clones*1
LD30	15-013	No clones*2
LD31	15-013	31-06, -09, -10, -26, -30, -31, -35, -38

Experiments with clones that failed during targeted integration of GFP in the parallel experiment were discontinued after clone analysis and showed either low cloning efficiency and lower expression level than obtained in the mini pool (24-1724 and 24-1797) or homogenous expression level (24-3043) indicating genetic instability and a homogenous mini pool respectively. Both instable clones and homogenous mini pools that do not survive double selection after targeting are not suitable host cell clones and were thus excluded from additional experiments. Among subclones of clone 24-0508, 25-004 had the highest activity of SEAP in cell culture supernatant and was tested in gene targeting experiments along with 3 other subclones to ensure that the right clone was chosen.

4.2.2.3 Productivity of secreted alkaline phosphatase

The screening of more than 4000 clones lead to a productivity that was increased 3.5-fold compared to 01C090 (Figure 25).

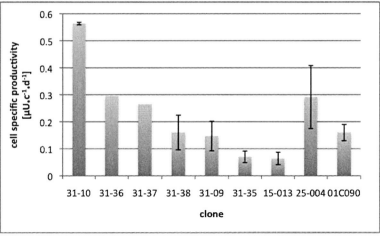

Figure 25: Cell specific productivity of selected CEMAX host cell clones. Subclone 31-10 showed a nearly 9-fold higher productivity of hSEAP compared to the initial pool 15-013. Productivity was increased 3.5-fold compared to the best clone of the first cell line generation. Error bars indicate either deviation between duplicate experiments or variations during cultivation.

From this analysis it was evident that mini pool 15-013 was a heterogeneous population of several subclones. Clone 31-10, the clone with the highest productivity, had a nearly 9-fold higher cell specific productivity than 15-013.

This result confirmed the existence of an outperformer clone in the heterogeneous mini pool 15-013. The expression level of GFP after gene targeting, which seemed comparable to GFP producer cells derived from the first host cell generation outperformer clones, supported this finding.

4.2.2.4 Survey of site-specific and functional integration

The applicability of clones for gene replacement at the target site was analysed in cotransfection experiments with the GFP replacement vector and the I-SceI expression plasmid. These experiments were performed with modifications to the established protocol to detect both end joining events and targeting events. These modifications were the selection with G418 and subsequent selection of resistant mini pools with Zeocin. Selection was performed in serum-supplemented medium in 96-well microplates for better reproducibility.

Several clones could be excluded from further experiments after this initial test either due to absolutely no growth of resistant cells after transfection (14-019, 24-1797, 24-2995 and 24-3043) or outgrowth of few resistant non-producer cells after G418 selection that proved sensitivity to Zeocin (15-152, 15-162, 24-1724).

Results from these clones were in contrast to host cell clones 24-0508, 15-013 and subclones that were successfully tested in gene targeting experiments. With these clones, which were used in additional experiments, it was possible to select double resistant and GFP positive cells. Of 24-0508's subclones 25-004 worked best in experiments with GFP and was chosen for further experiments. Several subclones of 15-013 were also capable for targeted integration and were analysed in more detail. Fluorescence microscopy was not suitable to detect differences in fluorescence intensity of excited GFP from producer cells derived of the first and second series of host cell clones. For the purpose of comparison and quantification of expression level genes encoding secreted proteins that are more in the focus of recombinant protein expression were used in further studies (see 4.3.3).

The effect of the tag vector modification was studied in a control experiment with mini pool 24-508, which was the parental mini pool of clone 25-004.

Resistant cells occurred at a frequency of at least 2.5·10⁻⁴ events per transfected cell after transfection with 8 µg of the I-SceI expression vector followed by G418 selection. This frequency was roughly the same as obtained with host cell clone 01C090. This was surprisingly in the context of the modification that was made in the tag vector. End joining required approximately 340 bp excessive resection of the transcriptional isolator. This was a 16-fold increase in comparison to previously described resections in the range of 1 bp to 21 bp [95, 117]. In addition to this, the right hand product of the I-SceI cleavage reaction exhibits high affinity to the endonuclease [118] and might thus interfere with processing of the DNA end.

4.2.2.5 Analysis of the integration pattern of the tag vector

Southern blot analysis was performed to characterise the integration pattern of the tag vector and to detect differences between mini pools and subclones derived thereof. Mini pool 15-013 was of special interest in this context. The southern blot analysis of selected clones derived from CV063 transfections is depicted in Figure 27. The corresponding map of a hypothetical single copy integration of the tag vector is shown in Figure 26.

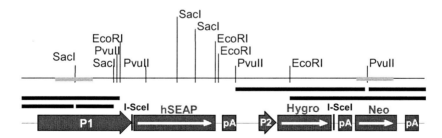

Figure 26: Restriction sites in the tag vector CV063 integrated as single copy integration.
Black bars indicate fragments that are detectable with the probes (grey bars). These are from top to bottom restriction fragments from PvuII, EcoRI, and SacI. Minimum fragment sizes are with the probe corresponding to the CMV promoter: EcoRI: >1883 bp; SacI: >1024 bp + 728 bp, and with the probe for detection of neo sequences: PvuII: >888 bp + 2415 bp; EcoRI: >2257 bp; SacI: >4060 bp. P1: enhanced CMV promoter with intron A; P2: SV40 early promoter.

Host cell clone 25-004 contains a single copy integration of the tag vector according to analysis with the CMV probe (Figure 27 B). One band from the

EcoRI restriction and one band plus the internal control fragment from the SacI restriction were detected. But two and three fragments detected with the probe specific for neo sequences in genomic DNA treated with SacI and PvuII indicated two copies of this region in host cell genome (Figure 27 A). An analysis of this clone in combination with targeted clones showed that sizes of these fragments kept constant or were altered specifically during recombination. This analysis is described in more detail in case study II (see section 4.3.3.3). These additional copies of the neo region did not interfere with the generation of producer cells. This was consistent with results obtained with host cell clone 01C090 that also contained two copies of the neo region in addition to three copies of the CMV region (see Figure 21).

Mini pool 15-013 and subclones were distinguishable by specific productivity and in Southern blot analysis. Several fragments detected in genomic DNA of 15-013 were found separated in four different subclone populations with increased band intensity. This indicated a mixed mini pool population composed of at least four subclones. The fragment sizes of subpopulation A to D are summarised in Table 6. The analysis of subpopulation C and D is not included in the blots in Figure 27. Differences in the analysis with the CMV specific probe were not detectable due to low resolution of high molecular weight fragments. Subpopulation B and D exhibited fragment patterns that indicate an intact single copy integration of the tag vector CV063. Subpopulation A showed bands from EcoRI and SacI treated DNA on the neo blot that might be derived of two fragments of similar size. Single copy integration was not assured with this analysis. Nevertheless, this did not prohibit expression of GFP after integration via the replacement vector.

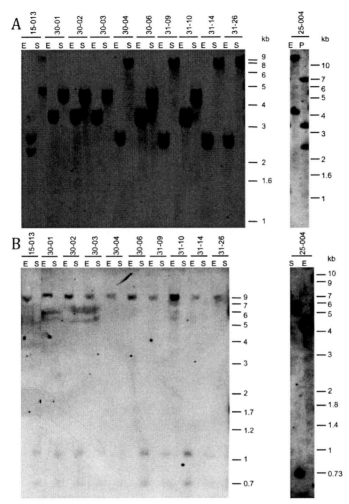

Figure 27: Southern blot analysis of the second series of CEMAX host cells.
Fragments containing sequences of the tag vector were hybridised with a neo (A) or a CMV specific probe (B). Genomic DNA was treated with EcoRI (E), SacI (S), or PvuII (P) for the analysis. Minimum fragment sizes are with the probe for detection of neo sequences for EcoRI 2257 bp, for SacI 4060 bp, and for PvuII: 888 bp plus an internal fragment of 2415 bp. With the probe complementary to the CMV promoter these were for EcoRI 1883 bp, and for SacI 1024 bp plus an internal control fragment of 728 bp. Fragment sizes deduced from a molecular weight standard are shown on the right side of the blots.

Table 6: Categories of subclones from 15-013 detected by Southern Blot analysis.
*1: was not analysed.

Group	Fragment size in kb					Examples
	Neo probe			CMV probe		
	BglII	EcoRI	SacI	EcoRI	SacI	
A	8.5	3.5	4.5	9	1.1+0.7	31-06, 31-10
B	>10	2.8	9	9	1.1+0.7	31-09, 31-35, 31-26
C	7+8	*1	*1	9	1.1+0.7	31-36
D	8	2.5	5.2	9	1.1+0.7	31-38

4.3 Generation of producer cells using targeted integration

Generation of producer cells by targeted integration was the central aspect of this work. After generating CEMAX host cell clones and assessing functionality by targeted integration of the GFP gene, genes encoding secreted model proteins were used for detailed characterisation of the expression level. Furthermore, reproducibility of productivity for similar proteins, the effects of the protein design on productivity after targeted integration, and the comparability of growth characteristics between host cells and producer cells were of interest.

Experiences from first gene targeting experiments with the GFP replacement vector from the assessment of clones for applicability of gene targeting were used to develop and optimise protocols for generation of producer cells (section 4.3.1). These protocols were subsequently applied for generation of producer cells for secreted proteins from the first (section 4.3.2) and second series (section 4.3.3) of host cell clones. Case study I (section 4.3.2.3) describes a possibility to implement the expression system in the field of recombinant protein production. Protein material of three isoforms of a potential immune suppressor for early development phases was produced from stable cells in this example. Case study II describes the development and characterisation of producer cells including confirmation of homologous recombination at the molecular level. Production of antibodies from CEMAX cells and characterisation of production stability are described at the end of this chapter.

Secreted proteins that were expressed after targeted integration were fusion proteins and antibodies. The biopharmaceutical product candidate CRB-15 is a Fc fusion protein that has four potential N-glycosylation sites per monomer. It prevents signal transduction in immunogenic processes by blocking the cytokine receptor for IL-15. Three isoforms of a potential immune suppressor exemplified a second biopharmaceutical product candidate. This proprietary fusion protein was a soluble version of a cell surface protein found on a wide range of immune cells. It was composed of an active domain fused to the Fc region of human immunoglobin G (IgG). All three variants of this fusion protein have two potential N-glycosylation sites per monomer. A third Fc fusion protein was expressed in CEMAX cells, which was a proprietary soluble selectin. Besides these fusion proteins two recombinant antibodies, ATROSAB and a second proprietary antibody, were used to evaluate the expression system. ATROSAB is a humanised antibody specifically binding the tumor necrosis factor (TNF) receptor 1 and is a potential biopharmaceutical for treatment of acute inflammatory diseases.

4.3.1 Development of protocols for gene targeting

Optimised and standardised protocols were essential for reproducible generation of producer cells. These protocols for gene targeting were developed with host cell clones of the first series using the GFP control replacement vector cotransfected with the I-SceI expression vector pCLS0197. The aim was to develop protocols for the serum-free selection of producer clones, a selection strategy for elimination of non-producer cells, and an optimisation of the transfection parameters DNA amount and ratio between the two plasmids.

4.3.1.1 Serum-free selection

As known from literature, the frequency for targeted integration via DSB induced homologous recombination was expected to be in the range of one event in 10^5 to 10^6 treated cells [61, 73, 105]. These rare events can be selected with the classical approach in serum supplemented culture of adherent cells, where dead cells are eliminated by removal of spent medium. Things became more difficult for selection in suspension culture where dead cells remain relatively stable for several days and are most efficiently removed by overgrowth of resistant cells and repeated dilution of the culture (see Figure 16 for illustration of removal of dead cells). Nevertheless, serum-free culture should also be applied during cell line development to establish a regulatory friendly process without the need of adaptation of producer cells to serum-free medium after selection.

First experiments were performed in serum containing culture in 150 cm^2 tissue culture dishes inoculated with $2*10^6$ transfected cells. Gene targeting events were subsequently selected for activation of the neo marker as colonies in selective medium. Temporary cultivation in serum containing medium for selection proved applicable. The relative ease of return to growth in serum-free suspension culture after temporary cultivation in serum containing medium was consistent with previous reports [119, 120].

Selection of cells, which have undergone site-specific integration, in serum-free culture was facilitated by the cultivation and selection of transfected cells in mini pools. Therefore the inoculation density of the serum approach in culture dishes was back calculated to 96-wells. This mini pool approach allowed increasing the fraction of positive cells from approximately 1 targeted integration event in 10^6 treated cells to 1 in 4200 cells in certain wells or mini pools, respectively. Outgrowth of cells that have undergone homologous recombination with the replacement vector was thus supported by the fact that the majority of negative cells was dispensed into other wells and did not interfere with selection. Five 96-well tissue culture test plates were used for selection of $2*10^6$ transfected cells at an inoculation density of roughly 4200 cells per well in the serum-free approach.

4.3.1.2 Selection strategy

First gene targeting experiments have shown that it was not sufficient to use the neo marker as solely selection means owing to a contamination with non-producer cells after selection. As discussed above, possible reasons for that might have been activation of the marker through end joining with linkage to the CMV promoter and generation of a start codon. As a second possibility for resistant non-producers was "one-sided" recombination (see section 4.2.1.3 and 4.2.2.4 for discussion).

To avoid incomplete selection, the tag vector was modified for the second cell line generation in order to reduce neo resistance due to illegitimate end joining. On the other hand, the zeo gene, which is linked to the GOI via an IRES, was incorporated in the replacement cassette to address the problem of "one-sided" recombination. This allowed selection of correct recombination events with markers on both ends of the replacement cassette that got activated through recombination. The optimised selection strategy included selection for completion of the neo marker by means of G418 three days after transfection and selection with Zeocin another three days later. The initial three days allowed the intracellular expression of I-SceI, the introduction and repair of DSBs and subsequent expression of neo after activation of the gene via homologous recombination with the replacement vector. Zeocin selection was started three days later to avoid selection biased for promoter trap events. These promoter trap events occurred after transfection of the replacement vector at a frequency of $5.5 \cdot 10^{-5}$ per transfected cell when selection was initiated the day after transfection and conferred resistance to Zeocin due to transcriptional activation of the selection marker by integration into an active cellular gene.

4.3.1.3 Optimisation of transfection conditions

The optimisation of DNA amounts and proportion of replacement vector and I-SceI expression vector was performed in two steps with the CEMAX clone 07-022, the GFP control replacement vector CV062 and the I-SceI expression plasmid pCLS0197 in culture dishes with adherent cells. The amount of plasmid DNA was varied between 1 µg and 8 µg for each vector. 8 µg I-SceI expression vector cotransfected with 4µg GFP control replacement vector, which was equivalent to 0.7 pmol, allowed highest gene targeting frequencies and highest proportion of producer cells.

In combination with the optimised selection strategy, up to 94% of double resistant colonies showed GFP expression. This demonstrated that it was effective to include the Zeocin marker in the replacement vector design to avoid resistant non-producer cells. These transfection parameters were used subsequently for all gene targeting experiments and proved to give reproducible results.

4.3.2 Cell line generation 1 for recombinant protein expression

01C090, 07-022 and 08-018 were the first clones that facilitated high expression of GFP after gene targeting and were submitted to additional experiments in order to analyse the expression level of secreted proteins. Low producer clones were not analysed in these experiments.

4.3.2.1 Host cell clones 07-022 and 08-018

Host cell clones 07-022 and 08-018 demonstrated the difficulty of achieving clones suitable for gene targeting, although gene replacement experiments with the GFP reporter were successful. 07-022 proved to be not reliable in gene targeting experiments focused on producing secreted proteins. This was consistent with the instability of SEAP productivity (see Figure 18) and the Southern blot analysis indicating an incomplete integration of the tag vector (see section 4.2.1.4). These were the reasons not to use this clone as a CEMAX host cell in terms of recombinant protein production.

Host cell candidate 08-018 was also not selected as host cell clone for in depth characterisation of the production of secreted proteins. This clone showed a lower productivity than 01C090 and low robustness of targeted integration in four targeting experiments with genes encoding fusion proteins.

4.3.2.2 Host cell clone 01C090

CEMAX host cell 01C090, which was monoclonal as deduced from productivity and Southern blot analysis, was the clone of cell line generation 1 that was used most extensively in gene targeting experiments. Except for a relatively high fraction of double resistant non-producer cells after double selection, this clone was used successful and reliable in several experiments. The cell specific productivity after targeted integration was between 1.6 $pg \cdot c^{-1} \cdot d^{-1}$ (IgG) and 2.7 $pg \cdot c^{-1} \cdot d^{-1}$ (fusion protein). Case study I (section 4.3.2.3) exemplifies the use of clone 01C090 and the virtue of the CEMAX system for rapid production of protein material from stable cell lines. In this particular study 50 mg of three isoforms of an antibody fusion protein had to be produced within 9 weeks for characterisation of these variants in *in vitro* studies. Table 7 gives an overview of productivities achieved with three different protein products.

Table 7: Cell specific productivity of host cell clone 01C090 after gene targeting. Values for productivity were determined either in batch or fed-batch experiments for the potential immune suppressor, or in 6-well assays in case of the proprietary IgG and CRB-15.

Gene product	Productivity
Potential immune suppressor isoform A	2.7 ± 0.1 pg·c^{-1}·d^{-1}
Potential immune suppressor isoform B	2.6 ± 0.1 pg·c^{-1}·d^{-1}
Potential immune suppressor isoform C	2.7 ± 0.2 pg·c^{-1}·d^{-1}
CRB-15	1.9 ± 0.1 pg·c^{-1}·d^{-1}
Proprietary IgG	1.6 ± 0.8 pg·c^{-1}·d^{-1}

With the three isoforms of the potential immune suppressor fusion protein it could be demonstrated that the expression level after gene targeting is nearly identical for highly similar proteins, as it could be assumed from theory. CRB-15, which contains two N-glycosylation sites more per monomer, is produced at a 30% reduced rate. This is probably owing to limitations in the secretion machinery of the producer cell [46]. As discussed below (section 4.3.5), this clone was not the best choice for antibody expression. Although antibodies are produced at higher rates than fusion proteins with other host cells, 01C090 showed low productivity and low recombination frequencies with this kind of protein.

4.3.2.3 Case study I: rapid production of protein material from stable cells

This case study gives an example how the CEMAX system can be implemented in the field of recombinant protein production. CEMAX host cell clone 01C090 was chosen for the small-scale production of three isoforms of a potential immune suppressor harbouring point mutations. Stable producer cell lines for production of additional protein material in a later development phase should be generated without huge clone screening effort. The first product had to be available within approximately 8 weeks after transfection, which is equivalent to 9 weeks including cloning of the replacement vector, in a quantity of 50 mg after affinity purification using Protein A. Additional material of one of the pharmaceutical product candidates had to be produced with the initial cell stock subsequently to evaluation and characterisation of the three isoforms.

Figure 28 illustrates the timescale and essential steps for the production of the three proteins. The coding sequences of the three proteins were cloned into replacement vectors by insertion into the cloning site. Meanwhile CEMAX host cells were recovered from a frozen cell stock and were then cotransfected with the modified replacement vectors and the I-SceI expression plasmid. Stable producer cells were selected and expanded as pools of a few clones to accelerate cell expansion. In this study 800 ml scale was sufficient for production. Nevertheless, the amount of product can be increased upon demand due to the scalability of the process based on the stable producer cell line. One purification step by protein A affinity chromatography was performed after clarification.

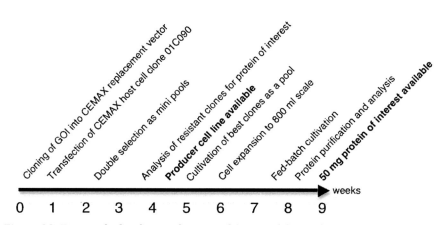

Figure 28: Timescale for the production of 50 mg of the Fc fusion protein.

Producer cells from three independent cotransfections of CEMAX host cell clone 01C090 were selected by means of G418 and Zeocin for 24 days. Figure 29 shows the result of the analysis of 96-well supernatant for isoform A of the drug candidate as an example. Samples were taken from 41 wells that showed signs of cell growth and, as a negative control, one well that did not. Among these 41 wells eight contained quantifiable amount of the fusion protein. The 4 clones with the highest product concentration were pooled for cell expansion.

Although host cell clone 01C090 has proven to be useful for reproducible and robust generation of producer cells, this clone showed a high degree of non-producer cells that failed to be eliminated during double selection in 96-well scale. This was consistent in several experiments and resulted in relative high numbers of clones that had to be analysed for the protein of interest. As discussed above (see 4.3.1.2) this might be a result of a combination of promoter trap activation of zeo and end joining activation of the neo marker conferring double resistance.

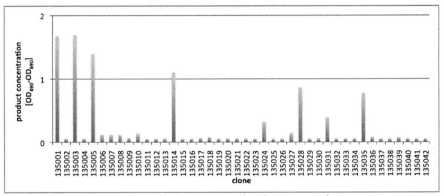

Figure 29: Analysis of 96-well supernatant for the potential immune suppressor isoform A.

Host cell clone 01C090 was used for gene targeting. Cells were selected with G418 and Zeocin until day 24 after transfection and were then inspected for growth of resistant cells by microscopy. The product of interested was detected by a human IgG specific ELISA and raw data of the measurement were used for the clone decision. Four clones (135001, 135003, 135005 and 135014) were cultivated as a pool for faster cell expansion in 12-well scale.

Starting with a 0.5 ml culture of pooled cells from 4 microplate wells the culture was expanded to 800 ml scale for production of the protein material. Figure 30 shows culture progression from expansion and production phase. Viability of the culture was low in the beginning, which was consistent with the other two protein isoforms. Through a phase of low cell density after transfer to spinner flask viability dropped to approximately 85% but recovered as higher cell densities were achieved. This was probably a result of attachment of viable cells to the glass wall as presumed from other experiments with 14-CHO-S cells. Low cell densities were avoided in subsequent expansion of producer cells for the isoforms B and C and viability was constantly high after initial recovery.

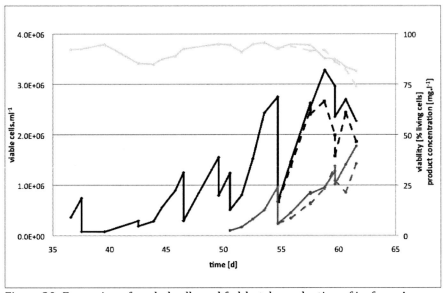

Figure 30: Expansion of pooled cells and fed-batch production of isoform A.
The pool of 4 clones was cultivated in 12-well and 6-well scale between day 29 and day 34 after start of the project. Viability in the 6-well was at 64%. Cells were then transferred into T25 and were submitted to spinner culture on day 37 with subsequently further increasing the culture scale to 800 ml. Two fed-batch experiments were inoculated from this culture on day 55 in the same scale. Culture 1 (continuous lines) was fed on day 2.5 and 4.5 of the fed-batch experiment as well as culture 2 (dashed lines). Cells were cultivated in BD CHO medium. The light grey lines show progression of viability. Two data in viable cell density (black line) at one point in time, connected through a vertical line, indicate dilution of cells. The dark grey lines represent product concentration.

The cultures were fed at glucose levels <3 g·l^{-1} in the production phase without any optimisation of the feeding protocol using standard feeding options. With the first feeding 2 g·l^{-1} glucose and 3 g·l^{-1} soy hydrolysate were added to the culture on day 2.5 of the fed-batch experiment. The second feeding was performed another two days later with 3 g·l^{-1} soy hydrolysate and addition of 170 ml medium to reach final process volume of 800 ml. Culture 1 reached a product concentration of 44 mg·l^{-1} while culture 2 reached 36 mg·l^{-1} which was consistent with a cell specific productivity of 2.7 pg·c^{-1}·d^{-1}. The small difference in concentration of the glycoprotein in the two spinner flasks was based on higher cell density and approximately 20% higher cumulative volumetric cell-days in culture 1. A biologic production capacity of 1.6·10^7 cell-days·ml^{-1} achieved in this not optimised experiment leaves much room for improvement in both maximum cell densities and culture duration. This could be achieved by assessing alternative culture media, especially the chemically defined medium CDM3 that was evaluated in a later section of this work (see section 4.4). Optimisation of the fed-batch strategy would then be the second step in optimising the process yield. In this experiment the first feed was added at nearly maximum cell density, which could have been too late for optimal culture performance.

Together with remaining cell suspension from the inoculum culture approximately 2 l cell culture supernatant were submitted to the production group for purification and analysis. This was equivalent to 71 mg protein of interest as summarised in Table 8.

Table 8: Harvest of isoform A of the potential immune suppressor.
In total 71 mg were harvested and 64 mg recovered after purification via protein A affinity chromatography from three batches of cell culture harvest.

Culture	Volume	Product concentration
Inoculum culture	450 ml	24 mg·l^{-1}
Fed-batch 1	750 ml	44 mg·l^{-1}
Fed-batch 2	750 ml	36 mg·l^{-1}

64 mg of isoform A as determined by bicinchoninic acid (BCA) assay were recovered and analysed by isoelectric focusing (IEF) among other techniques. As communicated by the colleagues, shifts in the isoform distribution in IEF were consistent with the theoretic isoelectric points (pI) of the protein variants. This indicated that the producer cells derived of a particular host cell process the protein products in a similar way, although more precise methods are necessary to produce direct evidence for this.

For comparison with the conventional method, a stable bulk pool from cells transfected with the expression vector of isoform A was selected in parallel to the strategy described above. The bulk pool was selected for 3 to 4 weeks and exhibited 6.7-fold lower cell specific productivity compared to the CEMAX producer cells. The product concentration did not surpass 3.7 mg·l^{-1}. Additionally, productivity decreased with prolonged cultivation time indicating instability and the approach was discontinued. Protein isoforms B and C were then produced with the CEMAX host cell 01C090 after feasibility was shown with this case study.

As described above, 64 mg of isoform A of the potential immune suppressor fusion protein were produced within 9 weeks including elementary steps as cloning the gene of interest, selection of stable producers, fed-batch production, protein A purification and analysis of the gene product. Short process times through minimisation of the clone screening process by targeted integration combined with serum-free cultivation demonstrate the virtue of the expression system for production and characterisation of several protein isoforms in early development phases. Proteins can be produced highly reproducible from a cellular system that is identical except for the coding sequence of the GOI. The possibility to produce comparable batches of protein material from frozen cells in a later project stages in any production scale is an advantage compared to transient expression systems with restricted possibilities for scale-up. The production group repeated the strategy successful for isoforms B and C afterwards. After characterisation of the three point-mutated variants, isoform C was chosen for additional experiments. Protein material therefore was produced in a 10 l wave bioreactor after thawing of stable producer cells from the initial experiment. The experiments demonstrated the capability for cultivation of this cell line in this reactor type.

A good alternative to host cell clone 01C090 would have been the use of clone 25-004. After this project was finished, a product concentration of 117 mg·l^{-1} was achieved in a batch experiment with isoform A of the fusion protein produced in 25-004 (see Figure 34). This would be an alternative for increasing the yield and shortening delivery times of the process. With an additional subculture step prior the batch or fed-batch experiment it would be possible to produce 1 g protein from a 10 l process in less than 10 weeks from start of the project.

4.3.3 Site-specific integration with the 2nd generation of host cell lines

The second, improved series of CEMAX host cell clones was developed to reduce the probability of neo activation through end joining after cleavage of the genomic tag with I-SceI. Out of theses, clone 25-004, and mini pool 15-013 and subclones were tested successful for targeted integration of genes for secreted protein products. The results are described below.

4.3.3.1 Mini pool 15-013 and subclones

From previous experiments it was deduced that mini pool 15-013 was composed of at least four different subclones (see section 4.2.2.5). This was consistent with gene targeting experiments performed with this mini pool. Producer cells exhibited different characteristics owing to the fact that they were derived of different subclone populations. This was shown for example after cotransfection with the replacement vector harbouring the gene of isoform A of the potential immune suppressor fusion protein. Three producer clones (202010, 202011, and 202027) were selected and subsequently analysed in more detail. The glycoprotein was produced at a rate of approximately 10 pg$\cdot$c$^{-1}\cdot$d^{-1} from this improved host cell series after replacement of the reporter cassette (Figure 31).

Although producer clones 202010 and 202027 produced the fusion protein at a similar high rate, these clones showed different growth characteristics and might thus be derived of different subpopulations. Maximum viable cell density in batch experiments in chemical defined medium CDM3, for example, was 7.1 $\cdot 10^6$ cells$\cdot$ml^{-1} with clone 202010 and 11$\cdot 10^6$ cells$\cdot$ml^{-1} with clone 202027. The third clone, 202011, had a lower cell specific productivity of 6 pg$\cdot$c$^{-1}\cdot$d^{-1}. This different behaviour was supported by the results of Southern blot analysis depicted in Figure 32.

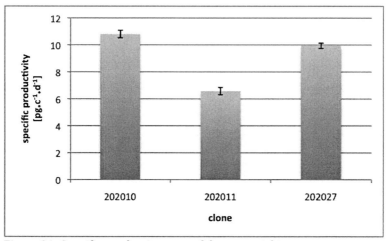

Figure 31: Specific production rate of the potential immune suppressor.
CEMAX host cell clone 15-013 was used for the gene targeting experiment. Error bars indicate deviation between duplicate experiments. Cells were cultivated in BD CHO medium in 6-well scale. Mean productivity of isoform A of the glycosylated fusion protein was calculated from a 3-day cultivation.

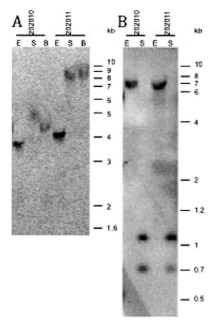

Figure 32: Southern blot analysis of producer clones 202010 and 202011. Fragments were detected with the neo specific probe (A) and CMV specific probe (B). Sizes of fragments were determined from a molecular weight standard and are indicated on the right side of the blot. Fragments of 202011 that were detected on the neo blot were larger than those of 202010. E: EcoRI, S: SacI, B: BglII.

Fragments detected with the neo probe from genomic DNA of 202011 were larger than those of 202010. This was valid for all three restriction endonuclease treatments indicating a different sequence background at the integration site (see Figure 32 for Southern blot and Table 9 for summary of fragment sizes). This points to the fact that the producer cells were derived of different subpopulations of 15-013 that have not been isolated in the subcloning experiment LD 31.

Table 9: Fragment sizes of 202010 and 202011 detected by Southern blot analysis.
Minimum sizes were calculated from the theoretical sequence after recombination and the plasmid sequence of the tag vector CV063. 202010 and 202011 were similar to host cell subpopulations B and D when only the results from detection of neo sequences were taken into account. But fragments of targeted clones detected with the CMV probe in EcoRI treated genomic DNA are not consistent with host cell clones analysed so far.

Group	Fragment size in kb				
	Neo probe			CMV probe	
	BglII	EcoRI	SacI	EcoRI	SacI
202010	4.2	3.7	4.7	7.5	1.1+0.7
202011	9	4	8	7.7	1.1+0.7
Minimum size host cell clone	>7.3	>2.3	>4.1	>1.9	1.0+0.7
Minimum size targeted clone	>4.0	>3.3	>4.1	>2.1	1.0+0.7

In addition to differences between the producer clones, fragments identified on the Southern blot with the CMV specific probe did not match with the integration pattern of known subclones. The fragments detected from EcoRI treated genomic DNA should have been enlarged by 0.2 kb after recombination as calculated from the theoretical sequence. But the apparent fragment size of 202010 and 202011 was at least 1.5 kb smaller than that of known host cell clones (Table 9 and Table 6 for comparison). The corresponding host cell clone of 202010, for example, would thus show a fragment with a size of 7.3 kb. But instead of fragments of this size, only 9 kb fragments were detected for all subclones analysed so far. This indicated that the subclones of 15-013, which were the host cell clones of producer clones 202010 and 202011, have not been isolated so far.

Reproducible generation of producer cells with high productivity would only be possible with an isolated host cell subclone. Otherwise a mixture of producer clones with different characteristics and a random chance to obtain a high producer like 202010 or 202027 would be the result. Taking into account a gene targeting frequency of $4 \cdot 10^{-6}$ events per transfected cell in this experiment and oligoclonality of 15-013, it was reasonable that the three clones were derived from different subclones of 15-013. They were thus not identical. Assuming an equal distribution of approximately six different subclones of 15-013 (four were isolated so far), and eight producer clones that arose out of $2 \cdot 10^6$ cotransfected cells, high producer clones like 202010 cannot be achieved reproducible from experiment to experiment. Hence, further work is necessary to isolate the corresponding subclone from mini pool 15-013. Using this subclone as host cell in gene targeting experiments would then ensure reproducible generation of producer cells like 202010 or 202027.

Producer clone 202027 was cultivated in batch and fed-batch experiments. A product concentration of 107 mg·l^{-1} was achieved in a batch experiment with BD CHO medium, which was also used for selection of targeted clones. With $2.4 \cdot 10^7$ cell-days·ml^{-1} this corresponded to a productivity of 4.5 pg·c^{-1}·d^{-1} at the end of the experiment. Nevertheless, productivity dropped from a peak productivity of 9.5 pg·c^{-1}·d^{-1} to less than 3 pg·c^{-1}·d^{-1} in the stationary growth phase due to limitation of nutrients. A maximum viable cell density of $3.9 \cdot 10^6$ cells·ml^{-1} was reached with this experimental set-up.

Afterwards productivity and cell growth were characterised in a fed-batch experiment in chemical defined medium CDM2 with the commercial feeding solution Cell boost 4 (Figure 33).

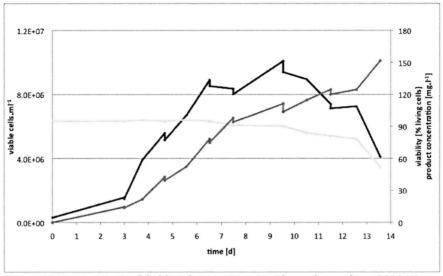

Figure 33: Non-optimised fed-batch experiment with producer clone 202027. The experiment was run in a controlled spinner bioreactor at pH 7.0 and 50% dissolved oxygen in chemically defined medium. Producer cells were cultivated in medium CDM2 and fed with Cell boost 4 on day 3, 5, 7, 8, 10 and 12. Glucose and glutamine were supplemented upon demand. Black line: viable cell density, light grey line: viability, dark grey line: product concentration. Two data at one point in time, connected through a vertical line, indicate dilution with feed medium.

Cells were recovered from a frozen cell stock and cultivated for 14 days prior start of the fed-batch experiment. Concentration of the potential immune suppressor Fc fusion protein could be increased to 152 mg·l⁻¹. Nevertheless, cell specific productivity decreased to 2.1 pg·c⁻¹·d⁻¹ at the end of the experiment. The peak productivity of 6.3 pg·c⁻¹·d⁻¹ was calculated in the beginning of the experiment and was approximately 30% lower compared to the cultivation in BD CHO medium. Decreased cell specific productivity might be due to growth in chemical defined medium. Hydrolysates have shown to increase cell specific productivity of recombinant CHO cells drastically [121, 122]. The batch experiment was performed with BD CHO medium, which contained a high amount of hydrolysates. CDM2, the medium used in the fed-batch experiment, was fully chemically defined and was not optimised for this particular producer clone. Productivity decreased constantly through the course of the experiment indicating that the feeding solution was not suitable to deliver cellular demands in nutrients and supplements. Nevertheless the combination of chemically defined medium and feed solution prolonged cell viability and allowed cultivation at increased cell densities. The increase in yield was thus based on a more than 3-fold increase in cumulative volumetric cell-days to $7.2 \cdot 10^7$ cell-days$\cdot$ml^{-1}.

In order to minimise problems in terms of reproducibility, subclones of 15-013 were assessed in gene targeting reactions. At least one host cell clone of every subpopulation was tested in experiments with the CRB-15 replacement vector. Highest productivities were achieved with host cell clone 31-26 with 4.7 ± 0.6 pg$\cdot$c$^{-1}\cdot$d^{-1}. The productivity level of 202010 was not reached in these experiments. Additional attempts to isolate subclones failed due to the use of BD Select CHO medium that did not allow single cell cloning at all. As discussed above, the right subclone of 15-013 has to be isolated in future approaches.

Results from gene targeting experiments with mini pool 15-013 showed the potential of targeted integration to achieve productivities in the range of 10 pg$\cdot$c$^{-1}\cdot$d^{-1} for glycosylated fusion proteins. This was attained with much less effort compared to conventional clone screening approaches. Nevertheless, more work in host cell development is necessary to isolate a single cell derived subclone of 15-013 to achieve more reproducibility after gene targeting compared to 15-013. Recent improvements in cloning frequency (see 4.4.2) enabled subcloning at high frequencies, which might facilitate the isolation of subclones that could not be expanded from single cells during this work.

4.3.3.2 Host cell clone 25-004

CEMAX host cell clone 25-004 has been used successfully and reliably in various targeting experiments. Table 10 gives an overview of productivity obtained after targeted integration of product genes encoding fusion proteins and two humanised antibodies. Differences in productivities of different proteins produced in 25-004 were consistent with data of host cell clone 01C090, except for antibodies. IgG were expressed at rates 1.7- to 2.5-fold higher compared to glycosylated fusion proteins. As suggested earlier, events downstream of transcription might be limiting factors in protein expression [123]. Especially highly glycosylated fusion proteins require more complex posttranslational modifications than IgG. Thus, these fusion proteins are likely to be expressed at lower levels.

Table 10: Cell specific productivity from host cell clone 25-004 after site-specific integration.
Data for productivity of ATROSAB were kindly provided by Verena Lorenz.

Gene product	Productivity
CRB-15	2.2 ± 0.1 pg$\cdot$c$^{-1}\cdot$d^{-1}
Potential immune suppressor isoform A	2.9 ± 0.1 pg$\cdot$c$^{-1}\cdot$d^{-1}
Soluble selectin fusion protein	2.0 ± 0.1 pg$\cdot$c$^{-1}\cdot$d^{-1}
Proprietary IgG	4.1 ± 1.3 pg$\cdot$c$^{-1}\cdot$d^{-1}
ATROSAB	5 pg$\cdot$c$^{-1}\cdot$d^{-1}

Culture performance in BD CHO medium was exemplified by a batch experiment of cells derived of 25-004 producing isoform A of the potential immune suppressor fusion protein (Figure 34). A product concentration of 117 mg$\cdot$l^{-1} was achieved through high cumulative volumetric cell-days of $3.5\cdot10^7$ cell-days$\cdot$ml^{-1} attained in 12 days of culture. The productivity was consistent with the value stated in Table 10 obtained in the 6-well assay. This product concentration was achieved in a batch experiment and was 2.7 times higher than achieved with 01C090 in case study I for the same protein in fed-batch experiments (see 4.3.2.3). Development of a fed-batch strategy and the use of 25-004 as host cell would be a mean for increasing yield in approaches like described in the case study.

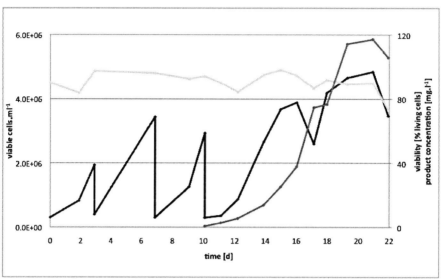

Figure 34: Cultivation and batch experiment of producer clone 22201.
Isoform A of the potential immune suppressor Fc fusion protein was produced from cells derived of host cell clone 25-004 after site-directed integration. The cells were cultivated in a T-flask until day 3 and then transferred into a spinner flask. Culture medium was BD CHO medium supplemented with 4 mM glutamine. The light grey line shows progression of viability, the dark grey line product concentration. Two data in viable cell density (black line) at one point in time, connected through a vertical line, indicate dilution of cells. The drop in viable cell density on day 17 was caused by a counting error.

A second example for culture performance including the comparison with the host cell is given in Figure 35. This batch experiment was performed with clone 571034 that produced a soluble selectin fusion protein. Host cell clone 25-004 (dashed lines) and the producer clone (continuous lines) were cultivated in parallel in chemically defined medium CDM3. As predicted from theory, the two cell lines performed similar regarding maximum viable cell density, batch duration and biological production capacity with values of $3.6 \cdot 10^7$ cells·d·ml^{-1} and $3.7 \cdot 10^7$ cells·d·ml^{-1}. This reproducibility of growth was also shown with other producer cells derived of 25-004. Consistent with the productivity shown in Table 10 was a concentration of the highly glycosylated fusion protein of 69 mg·l^{-1}.

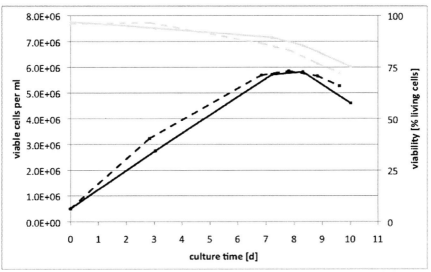

Figure 35: Batch experiment of producer clone 571034 compared with host cell clone 25-004.

Producer cells after targeted integration of the GOI (continuous lines) and CEMAX host cell clone 25-044 (dashed lines) were cultivated in CDM3 supplemented with 4 mM glutamine. The light grey lines show progression of viability, black lines show viable cell counts. Concentration of the fusion protein was 69 mg·l⁻¹ at day 10.

4.3.3.3 Case study II: characterisation of producer cells

This case study describes the characterisation of CEMAX producer cells and gives examples for confirmation of homologous recombination at the molecular level. The biopharmaceutical product candidate CRB-15 was produced after targeted integration of the coding sequence in host cell clone 25-004. Producer cells, which should be highly similar, were characterised for expression of CRB-15 and subsequently homologous recombination was confirmed by Southern blot analysis and with a specific PCR assay.

CEMAX host cells were cotransfected with a replacement vector containing the gene of the fusion protein and the I-SceI expression vector. Targeted cells were then selected in serum-free medium with the mini pool approach as described above (section 4.3.1). Positive clones were identified via generic ELISA in two rounds of measurement after three and four weeks of selection. The result of this analysis is illustrated in Figure 36. 46 clones showed initial cell growth. Out of these, 11 clones exhibited strong expression and 4 clones showed weak expression. No alkaline phosphatase activity was detectable in cell culture supernatant, indicating loss of the hSEAP reporter cassette as predicted from the recombination mechanism. Additionally, cells were no more resistant to hygromycin B, supporting the predicted loss of the reporter and selection cassette of CEMAX host cells. 31 clones, which showed cell growth, did not express the protein of interest. Most of those 31 clones did not grow after sub-culturing and were finally classified as false positive non-producers.

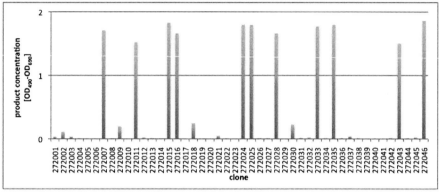

Figure 36: Analysis of 96-well supernatant for CRB-15 expression.
Host cell clone 25-004 was used for gene targeting. Cells were selected with G418 and Zeocin in 96-well mini pools and were then inspected for growth of resistant cells by microscopy. The product of interested was quantified by a human IgG specific ELISA and raw data of the measurement were used for the clone decision. Most non-producer clones could not be expanded beyond 24-well scale.

The 11 clones with strong expression of the product gene represented a gene targeting frequency of $5.5 \cdot 10^{-6}$ clones per transfected cell under serum-free conditions. This frequency is similar to results obtained with other host cell clones and product genes and is consistent with reports published earlier [61, 73, 105].

Clones that produced CRB-15 were expanded for further analysis including characterisation of productivity and verification of correct homologous recombination via Southern blot and PCR analysis of genomic DNA. Figure 37 shows cell specific productivity of the weak producer clone 272002 in addition to 7 randomly chosen clones.

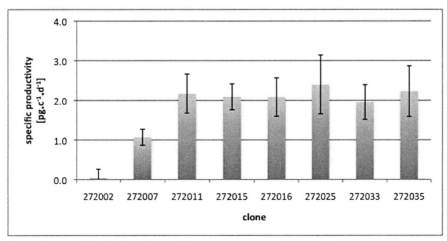

Figure 37: Analysis of cell specific productivity for CRB-15 with CEMAX host cell clone 25-004.
Error bars indicate deviation between measurements from cells cultivated for 1, 2 and 3 days respectively.

Six out of seven clones with high product concentration in the supernatant of the mini pools exhibited a comparable cell specific productivity of around 2 pg·c^{-1}·d^{-1}. This productivity was less than determined for other proteins expressed with the host cell clone 25-004. It might be caused by posttranscriptional bottlenecks due to four N-glycosylation sites per monomer. Nevertheless the production rate was comparable to the productivity of a cell line for production of clinical phase I material of CRB-15. This cell line was generated in an intensive screening process including cell sorting (unpublished observations, Cardion AG, Erkrath).

Clone 272002 could be expanded beyond 6-well scale although CRB-15 was barely detectable in cell culture supernatant of this clone. As deduced from the Southern blot analysis (Figure 39), this clone had undergone complex random integration events including integration of the I-SceI expression vector. These events may be linked with genetic instability and double resistance without or with low-level CRB-15 production.

Figure 38 shows maps of the target locus of an imaginary host cell clone (Figure 38 A) and the recombined locus with the CRB-15 replacement vector (Figure 38

B). The size of the internal fragment from PvuII treated DNA detected with the neo specific probe was decreased by 1.2 kb through recombination (*1 and *2 in Figure 38 and Figure 39).

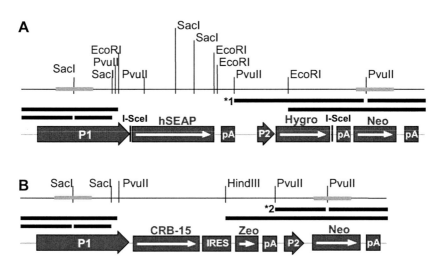

Figure 38: Map of the integrated tag vector and a recombined locus.
A: Tag vector CV063 integrated in the genome of the CEMAX host cell. B: Recombined locus after site-directed integration of the CRB-15 gene. Evident feature of gene targeting clones is an internal PvuII fragment detected with the neo probe, which has a reduced size of 1.2 kb compared to 2.4 kb from the host cell clone. Black bars indicate minimum fragment sizes calculated from known sequences that can be detected with the probes (grey bars). These are in A: PvuII, neo: >0.9 kb+2.4 kb; EcoRI, neo: >2.3 kb; EcoRI, CMV: >1.9 kb; SacI, CMV: >1.0 kb+0.7 kb. Minimum fragment sizes in B: PvuII, neo: >0.9 kb+1.2 kb; HindIII, neo: >2.9 kb; PvuII, CMV: >1.8 kb; SacI, CMV: >1.0 kb+0.7 kb. *1: internal fragment from PvuII restriction in host cells; *2: internal fragment from PvuII restriction after homologous recombination.

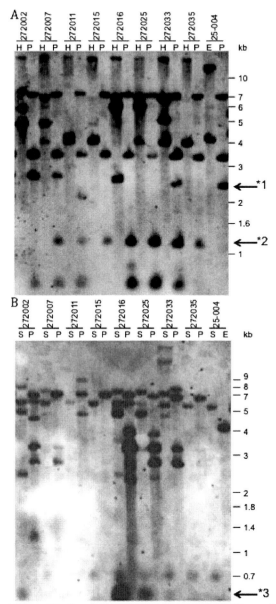

Figure 39: Southern blot analysis of CRB-15 producer cells.
Fragments were detected either with the neo probe (A) or the CMV probe (B).
E: EcoRI; H: HindIII; S: SacI; P: PvuII. *1: internal fragment from PvuII restriction in host cells; *2: internal fragment from PvuII restriction after homologous recombination; *3: 0.5 kb SacI fragment that can be ascribed to the I-SceI expression vector.

Southern blot analysis of clones obtained in gene targeting experiments revealed one problem that has to be solved in future to increase reliability of the system. The majority of clones had undergone random integration of either the replacement vector, or the I-SceI expression vector, or both in addition to targeted integration via homologous recombination. The hybridisation signal below the 0.7 kb control fragment in the SacI restriction on the CMV blot (*3 in Figure 39 B) may be ascribed to a 0.5 kb SacI fragment of the I-SceI expression vector which is complementary to the CMV probe. A typical example was clone 272016 with a strong signal at 0.5 kb along with other signals of high intensity. This indicated massive random integration of the cotransfected plasmids. Three other clones also showed this fragment (272002, 272025, and 272033). Integration of I-SceI expression vector, albeit at lower frequency, was reported earlier. Smih *et al.* found that 7.7% of targeted clones had undergone integration of the I-SceI expression vector [74]. Clones 272015 and 272035 showed no sign of random integration. These clones were seen as pure recombination clones without random integration of either replacement vector or I-SceI expression plasmid.

Gene targeting caused a characteristic fragment pattern. One feature of all targeted clones was a shift in the size of the internal PvuII fragment detected with the neo specific probe. The size of the internal fragment was 2.4 kb for the host cell clone (*1 in Figure 39 A). Since the second PvuII cleavage site was at a different position in the replacement cassette, a 1.2 kb smaller internal fragment was excised from recombined DNA (*2 in Figure 39 A). This change in fragment size was consistent with the theoretic sequence of the recombined locus. As expected, the size of the two large fragments from this restriction analysis (7 kb and 3.5 kb) did not change through recombination indicating that these fragments were derived of terminal sequences linked to genomic DNA that were not changed during recombination.

Another evidence for correct homologous recombination were unaltered fragment sizes from the SacI treated genomic DNA on the CMV blot. Both the terminal fragment (6 kb) and the internal fragment (0.7 kb) were not altered in size (Figure 39 B). The 6 kb fragment was derived of 1 kb tag vector sequence and 5 kb genomic DNA. This indicated that the integration site had not changed between host cell clone and producer cell. Genomic DNA of producer clones 272015 and 272035 was then used to produce direct evidence of homologous recombination at the molecular level by PCR (Figure 40).

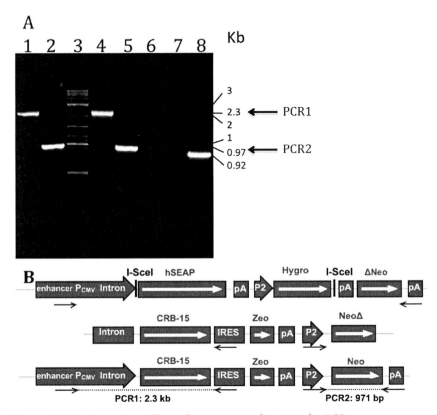

Figure 40: Confirmation of homologous recombination by PCR.
A: Analysis of PCR samples by agarose gel electrophoresis. Lane 1: 272015 PCR1, lane 2: 272015 PCR2, lane 3: 2-log DNA ladder (250 ng), lane 4: 272035 PCR1, lane 5: 272035 PCR2, lane 6: negative control PCR1 (tag vector), lane 7: negative control PCR2 (replacement vector), lane 8: positive control PCR2 (CV001). The samples were run on a 1.2% agarose gel and stained with ethidiumbromide. B: Primer binding sites and size of PCR fragments. From top to bottom: tag vector, replacement vector and recombined locus. Arrows indicate primer binding sites. PCR products from the recombined locus are represented by connected arrows. The size of the PCR fragment from the control is 54 bp smaller than that of the samples due to a shorter 3′ untranslated region of the neo in the control vector.

Therefore DNA fragments that span the recombination sites were amplified. The Assay allowed amplification of a DNA fragment only, if primer binding sites were combined on the modified locus after homologous recombination. Primer binding sites that demonstrate the specificity of the assay are shown in Figure 40 by arrows.

Both PCR assays of both clones gave a positive result and produced direct evidence of targeted integration of the CRB-15 expression cassette. A 2.3 kb DNA segment was amplified from PCR1 that covered the left hand recombination region and a 971 bp fragment from the right hand recombination region in PCR2. These sizes of amplified segments were consistent with the size calculated from the theoretic sequence after recombination.

This case study has shown that host cell clone 25-004 is suitable for production of CRB-15 at a level that was comparable to the corresponding cell line for production of clinical phase I material, which was derived of an intensive screening process. A productivity of 2.2 ± 0.1 pg·c^{-1}·d^{-1} was achieved with clones from gene targeting experiments that occur at a frequency of $5.5\cdot10^{-6}$ per treated host cell. Homologous recombination was verified subsequently with a PCR assay. Although Southern blot analysis revealed illegitimate integration events of replacement vector and I-SceI expression vector, a quarter of analysed clones did not contain detectable randomly integrated sequences. A possibility to overcome random integration would be to select against random integration of the replacement vector by a negative selection marker. Thymidine kinase of HSV is a well established negative selection marker for this purpose [57]. In addition, either purified enzyme electroporated into the cell [117] or transfection of mRNA encoding the endonuclease could substitute the I-SceI expression plasmid. Another possibility to reduce random integration of the I-SceI expression vector would be to incorporate the expression cassette in the backbone of the replacement vector. Moreover, this would allow fine-tuning of gene targeting frequency. Limitations in transfection efficiency as well as difficulties associated with delivery of two plasmids into the cell would be overcome.

4.3.4 Stability of production

Stability of protein expression is an important characteristic of production cells. It was analysed by cultivation for 61 cell generations in the absence of the selective compounds. Simulating a production run, this period covers the generation of master and working cell banks (10 generations each), expansion to production scale and production process (30 generations) including a backup of 10 cell generations. The stability was characterised during prolonged cultivation employing productivity measurements and batch experiments as well as flow cytometric analysis of intracellular product following immunofluorescent labelling.

Producer clone 571034 that was derived from mini pool selection after cotransfection of host cell clone 25-004 was chosen for this analysis. This clone secreted a soluble selectin fusion protein at a rate of 2.0 ± 0.1 pg$\cdot$c$^{-1}\cdot$d^{-1}. Figure 41 shows culture progression of cultivation in the absence of selective compounds compared to the reference culture with maintained selection pressure. A batch experiment of both cultures was performed on day 15 and on day 65 at the end of the experiment to assess culture performance. In the second batch experiment the test culture and reference culture reached similar biologic capacities for production of $4.8\cdot10^7$ and $4.9\cdot10^7$ cell-days$\cdot$ml^{-1}. This was a 30% to 35% increase compared to a batch experiment performed at the beginning of the study on day 15. The increased production capacity was based on higher growth rate and maximum viable cell densities and was consistent with increasing cell densities through continuing subcultivation. This observation was consistent with a recent report [124]. Besides this increase in growth rate and peak cell density, productivity between day 57 and 67 had decreased to 1.5 ± 0.1 and 1.6 ± 0.0 pg$\cdot$c$^{-1}\cdot$d^{-1} in the test and the reference culture. Compared to the initial value of 2.0 ± 0.1 pg$\cdot$c$^{-1}\cdot$d^{-1} this represents a 20% to 25% decrease. However, the decrease in cell specific productivity was compensated by the 30% to 35% increase in cumulative volumetric cell-days. Prentice *et al.* have also observed a loss in productivity in attempts of "bioreactor evolution", although this was not compensated by higher cumulative volumetric cell-days in their approach. After 5 iterative batch experiments in spinner flasks cell specific productivity was reduced by 45% [125]. Interestingly in the sense of experimental set-up was the fact that cell specific productivity remained more or less constant after iterative fed-batch experiments in a controlled bioreactor owing to different selective pressure in the reactor systems and during routine subculturing [125].

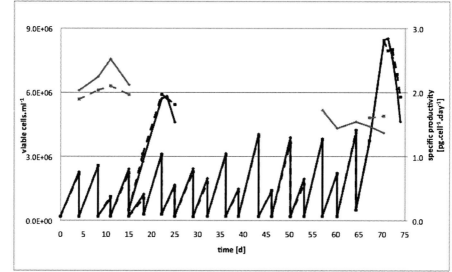

Figure 41: Prolonged cultivation of CEMAX producer clone 571034.
The test culture (continuous lines) was cultivated in the absence of selective compounds while reference culture (dashed lines) was cultivated in the presence of G418. Two data in viable cell density (black lines) at one point in time, connected through a vertical line, indicate dilution of cells. The grey lines show cell specific productivity. Cells were cultured in CDM3 supplemented with 4 mM glutamine and in the reference culture additionally 300 $\mu g \cdot ml^{-1}$ G418.

Flow cytometric analysis following immunofluorescent labelling of the intracellular product was performed as an additional tool to analyse stability of production. Results from this analysis (Figure 42) showed no change in the intracellular amount of the fusion protein as deduced from fluorescence intensity of the cell population indicating stability of protein production. Fluorescence intensity after 36 and 64 days in culture (green and blue) was identical to the intensity from the start of the experiment (black). The test and reference culture showed no differences. Instabilities, which would have been apparent from formation of subpopulations or shifts in fluorescence intensity, were not detectable.

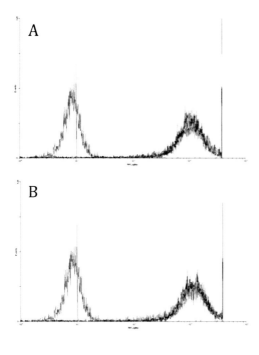

Figure 42: Histogram overlays after cytometric analysis. Intracellular product was stained with a Phycoerythrin (PE) labelled antibody and analysed on the Guava PCA. A: Test culture, cultivated in the absence of selective compounds. B: Reference culture, cultivated in the presence of G418. Red lines: negative control (25-CHO-S cells), black lines: 571034 at the start of the stability study, green lines: 571034 after 36 days in culture, blue lines: 571034 after 64 days in culture. Negative control and 571034 at the start of the experiment are identical in A and B.

The loss of productivity detected by assessing specific productivity during prolonged cultivation was thus either not detectable in the cytometric analysis, or was the effect of inaccurate determination of productivity and thus not real. Potential error sources arise through subcultivation, in cell counting, and determination of product concentration by ELISA with imprecisions of 10% to 30% each. Thus, accumulated errors could be the reason for the decreased productivity.

While instability of highly amplified antibody producing cells is often assumed to be a result of loss of genetic material, it may have different sources ranging from transcription, posttranscriptional processing, translation, posttranslational processing, to secretion [109]. A recent study demonstrates the susceptibility of production cells to loss of productivity. All out of 11 amplified high producer clones showed 54% to 97% loss of specific productivity after prolonged cultivation in the absence of MTX. In this case loss of productivity was consistent with significantly reduced mRNA levels [124]. In contrast to the susceptibility of highly amplified cell lines to loss of genetic material, protein expression from a single copy of a transgene should reduce the probability of instable expression.

Stability of protein expression was confirmed by cytometric analysis in this work standing in contrast to instability of highly amplified production cells. Although a slightly decreased cell specific productivity was detected during prolonged cultivation in an ELISA and cell counting based method, no changes of fluorescence intensity were detected in the cytometric analysis. Nevertheless, a change in growth characteristics occurred. Higher peak cell densities and higher cumulative volumetric cell-days were detected. This change compensated for the slightly reduced cell specific productivity observed in this experiment. A second cell line stability study was performed using 25-004 derived producer cells expressing an intracellular protein using cytometric analysis confirmed stability of production over a period of two months [126].

4.3.5 Antibody expression with CEMAX cells

Antibodies were used as model proteins for expression with the CEMAX system in addition to the fusion proteins described above. A prerequisite for antibody expression with CEMAX cells was a modification of the replacement vector. To allow expression of both heavy and light chain of IgG molecules a second expression cassette had to be inserted. This additional cassette contained the heavy chain gene and was inserted into the PsiI site within the replacement cassette (see Figure 14). The additional expression cassette was positioned downstream the regular expression cassette, which contained the light chain gene. The expression cassettes were positioned in the same direction of transcription (head-to-tail). In experiments described in this thesis the expression of the heavy chain was controlled by a promoter composed of the core promoter of the human elongation factor 1α gene [10] combined with 5′ untranslated region derived of human T-cell leukaemia virus [127] derived of pFUSE-Fc. With this modification the size of the replacement cassette increased to about 5.5 kb, which is 2.5 kb more than in the GFP control replacement vector.

Targeted integration of coding sequences for the proprietary IgG in host cell clone 01C090 caused problems. Recombination frequency was reduced to $1 \cdot 10^{-6}$

events per cotransfected cell resulting in lower reproducibility between experiments. Beside this, cell specific productivity of 1.6 ± 0.8 pg·c^{-1}·d^{-1} was below levels obtained for fusion proteins in previous experiments (see Table 7 for comparison).

Host cell clone 25-004 proved more suitable for expression of this protein class. Antibody producer cells could be generated after targeted integration at a frequency of $9\cdot10^{-6}$ events per transfected cell. This was one of the highest frequencies obtained from experiments with the serum-free selection strategy in this work. Figure 43 shows the distribution of productivity of 10 randomly chosen producer clones of the proprietary IgG1 antibody.

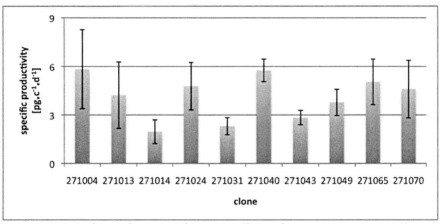

Figure 43: Specific productivity of CEMAX cells producing a proprietary IgG. Clone 25-004 was used as host cell for the gene targeting experiment. Error bars indicate deviation between measurements from cells cultivated for 1, 2 and 3 days respectively. Data were obtained in a 6-well assay in BD CHO medium.

The mean productivity of cells cultivated in BD CHO medium was calculated to 4.1 ± 1.3 pg·c^{-1}·d^{-1} with peak productivities of about 6 pg·c^{-1}·d^{-1}. This productivity was significantly higher compared to the production rate obtained for fusion proteins. These fusion proteins contained at least one N-glycosylation site more on each peptide chain and might thus have required more complex posttranslational processing. This might indicate limitations in the protein secretion machinery of CHO cells that are not known as professional secretors in their original tissue [42]. Variation of productivity between producer clones was higher with this antibody replacement vector compared to experiments where a single gene was inserted (see Figure 37 for comparison). This observation could have been a result of a reduced accuracy of recombination based on the increased size of the replacement cassette.

Additional work on expression of antibodies with CEMAX host cell clone 25-004 was performed at Celonic GmbH in Jülich. ATROSAB purified from CEMAX producer cells showed no differences in affinity for the TNF receptor 1 compared to the molecule expressed from stable CHO cells after random integration of an expression vector. A specific productivity of 5 $pg \cdot c^{-1} \cdot d^{-1}$ was consistent with the results achieved in this work.

In principal it has been shown that antibodies can be produced after targeted integration of the GOI via double-strand break induced homologous recombination. The size of the replacement cassette did not interfere with recombination frequency when host cell clone 25-004 was used for the experiments. Albeit productivity and gene targeting frequency were lower than usual with host cell clone 01C090. This indicated limitations in the process of targeted integration of larger replacement cassettes in this clone. A reason for this might be that gene targeting is a locus specific [57, 62] and thus a clone specific event. The increased size of the replacement cassette had an effect on accuracy of recombination. This was observed through higher variation of cell specific productivity between producer clones compared to experiments where only one expression cassette was inserted. Peak productivities of 6 $pg \cdot c^{-1} \cdot d^{-1}$ were achieved for the proprietary IgG1 antibody after targeted integration using a non-optimised medium. This was only 2.5-fold below the lower levels of productivity usually achieved in antibody expression from fully optimised production processes [43]. Improvements in productivity may be achieved, for example, by optimisation of basal media, by feed media optimisation [128], as well as culture parameters like pH [128]. Cell cycle arrest by reduced cultivation temperature, nutrient based proliferation control, and chemical approaches has shown to increase specific productivity [50, 129]. CHO based production cell lines usually require gene amplification to reach productivity levels sufficient for manufacturing. However, it has been reported that mouse myeloma cell line NS0 is capable of production of antibodies at productivities of up to 58 pg·c^{-1}·d^{-1} from a single copy of the transgene [130]. Such a cell line with high secretory potential from a single transgene copy could be an ideal host cell for the CEMAX system. But the major hindrance in the use of the NS0 cell system is its expression of sugar residues that are not present in humans and might be antigenic [131]. Nevertheless, targeted integration has the virtue that extensive clone screening and gene amplification are not necessary and expression of recombinant proteins from stable cell lines in short time frames is possible. Recovery of cells from a frozen cell bank allows repeated production batches with identical stable cells and free choice of cultivation strategy. This is not possible with transient expression where process strategy is limited to batch or fed-batch experiments.

4.4 Serum-free cultivation and cloning

Serum-free cell culture has become standard for production of biologics. Media improvements were one of the major driving factors for improved process yields through the past two decades. Serum removal is beneficial in terms of cost of goods, process performance and consistency as well as ease of protein purification. In addition regulatory concerns arose through the undefined character of serum including the risk of viral or prion protein contamination. Although removed from most production processes for recombinant proteins, serum is still used in cloning experiments during cell line development. Serum-free cloning has not been established as standard process in the development of recombinant cell lines.

At the beginning of this work the platform technology SEFEX (serum-free cloning and expression) was already established in the laboratory of Celonic. Based on the CHO-K1 derived cell line 14-CHO-S [132], SEFEX technology was successfully used for development of the first and second host cell line generation and serum-free selection of cells that had undergone targeted integration as well. Cloning efficiencies between 5% and 20% depending on experimental conditions were usually achieved in these experiments. Discontinuation of the serum-free medium BD CHO and replacement by BD Select CHO medium caused a drastic reduction in cloning efficiency to less than 0.5%. This became apparent during development of the third CEMAX host cell line generation (appendix 8.2). A short-term solution of this problem was the use of untransfected CHO cells as feeder cells. The feeder cells supported outgrowth of stable cells and were subsequently eliminated through selection pressure. This stimulated the outgrowth of clones to a cloning efficiency of 8%. Due to the potential risk of cross contamination a more sophisticated solution was needed and subsequently achieved by assessing the ability of commercial available media to support clonal growth as described in the next sections.

4.4.1 Media screening for serum-free cloning of 14-CHO-S cells

In contrast to BD CHO medium, which was used successful for serum-free limited dilution cloning, BD Select CHO medium proved to be not capable for single cell cloning experiments with 14-CHO-S cells. Hydrolysates from a different source in the new formulation were assumed to cause these problems. In order to solve this issue, 14-CHO-S cells were adapted to 21 serum-free media of commercial sources to overcome the limitations of BD Select CHO medium. Briefly, a batch experiment and a cloning experiment were performed after four subcultivation steps in the test media. Key parameters for the selection of the most suitable medium were growth characteristics in the batch experiment and, as an important factor for cell line development, the cloning efficiency.

In summary the chemically defined media CDM2, CDM3 and CDM6 permitted limited dilution cloning with acceptable cloning efficiencies in the range of 10% to 16%. Out of these media, CDM3 promoted the best growth characteristics as illustrated by a maximum viable cell density of $9*10^6$ cells$*$ml^{-1}. The cloning efficiency of 14-CHO-S cells in 13 media evaluated in this experiment is depicted in Figure 44. Except for CDM2, CDM3 and CDM6 no medium supported growth from single cells that was suitable for clone screening approaches what exemplifies the challenge of serum-free single cell cloning.

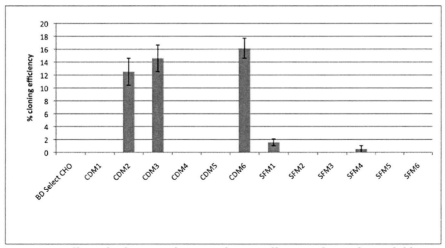

Figure 44: Effect of culture medium on cloning efficiency during limited dilution cloning.

Cells adapted to each test medium were inoculated into two 96-well microplates with 1 cell per well. Cell growth, detected by microscopy after cultivation for 14 days, was used for determination of cloning efficiency. Error bars show the deviation between duplicate plates. Cloning efficiency was calculated as the percentage of wells allowing outgrowth of cells per inoculated 96-well microplate.

Cell growth in CDM3 during adaptation phase and the batch experiment is illustrated in Figure 45. Compared to the reference medium (Figure 45 A) viable cell density was doubled from about $2 \cdot 10^6$ cells·ml^{-1} to about $4 \cdot 10^6$ cells·ml^{-1} three days after subcultivation in CDM3 (Figure 45 B). This would allow more effective cell expansion during cell line development or inoculum train. Furthermore the growth after the first subcultivation steps did not show the need of an adaptation phase. Higher viable cell density and a one day longer cell survival during the batch experiment in CDM3 resulted in a significant increase in cumulative volumetric cell-days from $1.8 \cdot 10^7$ cell-days·ml^{-1} to $4.0 \cdot 10^7$ cell-days·ml^{-1}.

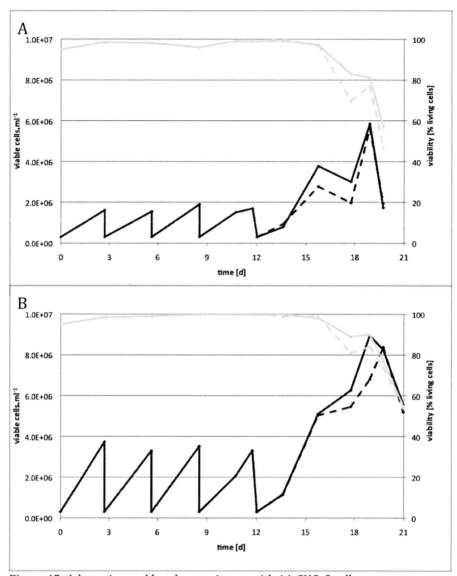

Figure 45: Adaptation and batch experiment with 14-CHO-S cells.
A: Reference culture with BD Select CHO medium. B: Test culture in CDM3. After the adaptation phase a limited dilution cloning experiment was performed on day 11 followed by a batch experiment run in duplicates on day 12. Two data in viable cell density (black lines) at one point in time, connected through a vertical line, indicate dilution of cells during the adaptation phase. Dashed lines represent the duplicate culture of the batch experiment. The light grey lines show progression of viability.

Table 11 compares key performance parameters of all media evaluated in this experiment. Media from one supplier (SFM7, SFM8, CDM7, CDM8, and CDM9) were excluded from further experiments after the adaptation phase due to formation of cell aggregates. These aggregates caused problems in cell counting, low cell densities and would especially interfere with single cell cloning. Without destruction of aggregates it would not be possible to generate a single cell derived clone. The cultures from the adaptation were continued after the third subcultivation to get a roughly idea of maximum viable cell densities and cumulative volumetric cell-days in batch experiments. Due to low cell densities and low viability during adaptation SFM9 and SFM10 were excluded from the experiment after two respectively four subcultivation steps. This observation could be based on the nutrient formulation of these two media, which was optimized for cultivation of hybridoma cell lines. Direct adaptation was also unsuitable for CDM5 and CDM10. The latter did not contain insulin or insulin like growth factor [133] what might have been the reason for a drop in viability and limited cell growth.

After assessing the key parameters listed in Table 11, the decision for a culture medium was drawn by calculating a score according to the matrix in Table 12. Therefore the best medium for each parameter got the full parameter score whereas other media got a fraction of it depending on the relative performance. The total score was the sum of all parameters (Figure 46).

The chemically defined media CDM1, CDM2 and CDM3 represented a substantial improvement compared to the reference medium but also compared to all other test media. The higher score of these three media was mainly based on higher viable cell densities combined with prolonged survival in the batch experiment resulting in higher cumulative volumetric cell-days. Taking feasibility for limited dilution cloning into account, CDM3 is a medium suitable for serum-free cell line development and as a superior basis for process development. An alternative to it is CDM2 with a slightly lower score. CDM6 was suitable for cloning experiments but showed significant disadvantage in cell cultivation due to low cell densities with this cell line.

Table 11: Key parameters of culture performance of 14-CHO-S cells in test media.
∗: values for cumulative cell-days and maximum viable cell density resulted from
a continuation of adaptation culture after subcultivation 3; n.d.: not determined.

Medium	cloning efficiency	cumulative cell-days	maximum viable cell density	growth rate $\mu_{day\ 1-3}$	batch duration	viability day 1-6
unit	%	cell-days·ml^{-1}	cells·ml^{-1}	1·d^{-1}	d	%
BD Select CHO	0.0	$1.8 \cdot 10^7$	$5.7 \cdot 10^6$	0.64	7.7	90
SFM1	1.6	$1.3 \cdot 10^7$	$2.1 \cdot 10^6$	0.52	8.9	95
CDM1	0.0	$3.6 \cdot 10^7$	$8.2 \cdot 10^6$	0.62	8.9	93
CDM2	12.5	$3.9 \cdot 10^7$	$6.8 \cdot 10^6$	0.79	8.9	94
CDM3	14.6	$4.0 \cdot 10^7$	$8.7 \cdot 10^6$	0.77	8.9	94
SFM2	0.0	$1.3 \cdot 10^7$	$2.0 \cdot 10^6$	0.46	9.5	96
CDM4	0.0	$1.7 \cdot 10^7$	$4.3 \cdot 10^6$	0.75	6.9	90
SFM3	0.0	$2.1 \cdot 10^7$	$4.3 \cdot 10^6$	0.72	7.8	91
CDM5	direct adaptation not successful					
SFM4	0.5	$1.6 \cdot 10^7$	$3.7 \cdot 10^6$	0.65	7.8	94
SFM5	0.0	$1.5 \cdot 10^7$	$4.0 \cdot 10^6$	0.67	6.9	85
SFM6	0.0	$1.8 \cdot 10^7$	$4.5 \cdot 10^6$	0.76	6.9	93
CDM6	16.1	$1.3 \cdot 10^7$	$2.3 \cdot 10^6$	0.53	8.9	93
CDM7	n.d.	$1.7 \cdot 10^7$	$3.9 \cdot 10^6$	formation of aggregates; ∗		
CDM8	n.d.	$2.2 \cdot 10^7$	$3.9 \cdot 10^6$	formation of aggregates; ∗		
CDM9	n.d.	$1.3 \cdot 10^7$	$1.8 \cdot 10^6$	formation of aggregates; ∗		
SFM7	n.d.	$2.0 \cdot 10^7$	$4.7 \cdot 10^6$	formation of aggregates; ∗		
SFM8	n.d.	$2.1 \cdot 10^7$	$3.8 \cdot 10^6$	formation of aggregates; ∗		
CDM10	direct adaptation not successful					
SFM9	direct adaptation not successful					
SFM10	direct adaptation not successful					
CDM11	n.d.	$2.6 \cdot 10^7$	$4.2 \cdot 10^6$	T75, 8% CO_2; ∗		

Table 12: Parameters and parameter score for rating of test media.
[*1]: data obtained from batch cultivation.

Parameter	Parameter score
Cloning efficiency	10
Cumulative volumetric cell-days [*1]	10
Maximum cell density [*1]	6
$\mu_{\text{day 1-3}}$ [*1]	8
Batch duration [*1]	7
Average viability day 1-6 [*1]	5

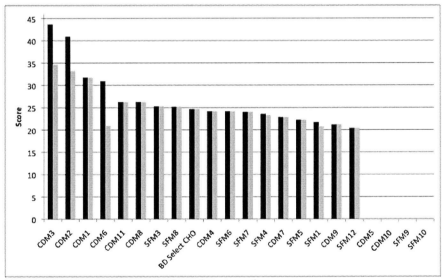

Figure 46: Rating of test media based on a performance score.
The maximal possible score was 46. Black bars: score including cloning experiment; grey bars: score without taking cloning efficiency into account. CDM11 was used in a T-flask at 8% CO_2.

Although production rates for a protein of interest were not included in the experimental design, CDM3 was chosen for further experiments. These included the generation of the serum-free growing CHO-K1 host cells 23-CHO-S and 25-CHO-S, which were derived by direct adaptation of 14-CHO-S and CHO-K1 cells to CDM3.

4.4.2 Media screening for improvement of serum-free cloning

Cloning efficiency is the limiting factor in clone screening approaches when a high number of clones has to be assessed. The cloning efficiency in CDM3 was in the range of 15% at inoculation densities of 1 cell per well. This means that over seventy 96-well microplates have to be inoculated to analyse 1000 clones during cell line development. To increase efficiency of limited dilution cloning, further development was focused on increasing the cloning efficiency. 25-CHO-S cells, which were directly adapted to growth in chemically defined medium from serum-dependent CHO-K1 cells, were subjected to temporary use of 14 test media in limited dilution cloning experiments.

Temporary use of the media was chosen due to the superior growth characteristics of 25-CHO-S cells in CDM3 and problems with some media including the tendency to form aggregates observed in the previous experiment (section 4.4.1). These media were SFM7, SFM8 and CDM7-9 owing to formation of aggregates, CDM6 due to low cell densities, and SFM11 that was developed for cloning experiments and not for routine cultivation. Based on the results of the prior experiment CDM10 was supplemented with insulin (10 mg·ml^{-1}). CDM12, SFM11 and SFM12 were not tested in the previous experiment.

Figure 47 shows the cloning efficiency obtained by temporary cultivation of 25-CHO-S cells in test media in two individual cloning experiments. Cloning efficiency was increased from 7±9% in CDM3 to 44±2% by the use of SFM12. SFM11, which was designed for cloning experiments, and CDM10 increased cloning efficiency to 15±12% and 31±6% respectively. An interesting feature of SFM11 was the ability to support adherent growth of cells in a serum-free formulation. Supplementation of CDM10 with insulin proved to be essential for single cell growth, although a high variation between the duplicate experiments was observed. This medium lacking insulin did not allow cloning of 25-CHO-S cells.

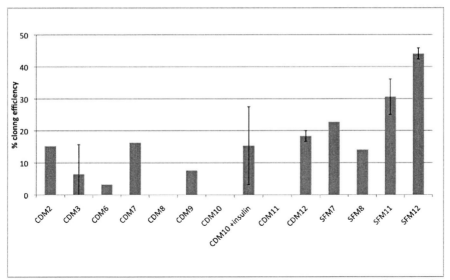

Figure 47: Cloning efficiency of 25-CHO-S cells with temporary use of test media. Cells were cultivated in CDM3 prior and after the cloning experiment. Error bars indicate deviation from two individual experiments with duplicate plates in the second experiment. CDM3 did not allow single cell cloning in the second experiment. Cloning efficiency was calculated as the percentage of wells allowing outgrowth of cells per inoculated 96-well microplate.

CDM3 did not allow single cell cloning in the second cloning experiment. This kind of variation from experiment to experiment was also observed with CDM3 in other projects and indicated that slight differences in experimental setup result in different success rates with this medium. Cloning efficiency in SFM12 remained constantly high between the two experiments. One critical parameter influencing cloning efficiency might be CO_2 and its effect on the pH of the media. Preincubation of CDM3 for adjustment of CO_2 and pH allowed outgrowth of cells at high frequencies, whereas frequent microscopic observation of cells or omitting the preincubation step resulted in either high variations from experiment to experiment, or even failure of cloning. Higher reproducibility with SFM12 might be due to the zwitterionic buffering agent HEPES that is included in this medium. Supplementation with HEPES buffer would thus be an approach for optimisation of other media for cell cloning applications in future.

25-CHO-S cells were subsequently adapted to CDM12, CDM10 supplemented with insulin, and SFM12 to assess performance in routine cultivation. All three media chosen for adaptation proved capable for cultivation of 25-CHO-S cells. Key parameters of culture performance are shown in Table 13. Potential biologic production capacity of 25-CHO-S cells cultivated in CDM3 was increased 1.8-fold to cumulative volumetric cell-days of $7.2 \cdot 10^7$ cell-days·ml^{-1} compared to the 14-CHO-S cell line cultivated in this medium (see Table 11 and Table 13). This might be an effect of direct adaptation of CHO-K1 cells instead of 14-CHO-S cells to the chemically defined medium.

Table 13: Key parameters of culture performance of 25-CHO-S cells in test media. Viability between day 1 to 6 of the batch experiment is not noted in this table since the value was 97% in all media.

Medium	cloning efficiency	cumulative volumetric cell-days	maximum viable cell density	growth rate $\mu_{day\ 1-3}$	batch duration
unit	%	cell-days·ml^{-1}	cells·ml^{-1}	$1 \cdot d^{-1}$	days
CDM3	4.7	$7.2 \cdot 10^7$	$10 \cdot 10^6$	1.00	11
CDM12	5.2	$3.0 \cdot 10^7$	$9.9 \cdot 10^6$	0.98	7
CDM10	28	$3.2 \cdot 10^7$	$4.5 \cdot 10^6$	0.62	13
SFM12	45	$2.3 \cdot 10^7$	$5.1 \cdot 10^6$	0.99	8

CDM3 provided the best culture performance in all parameters except for batch duration where CDM10 allowed culture progression for two additional days. Regarding maximum viable cell density, only CDM12 was comparable to CDM3 with nearly $10 \cdot 10^6$ cells·ml^{-1}. Due to the superiority of CDM3 regarding cell cultivation compared to the other test media it was decided to cultivate the cells in CDM3 and to use SFM12 temporary in limited dilution cloning experiments. This strategy increases cloning efficiency and robustness from experiment to experiment combined with superior growth promoting effect of CDM3 for routine cultivation and as a starting point for process optimisation.

It has been shown earlier that product yields and product characteristics are similar after transfections of preadapted CHO host cells and those derived from serum-containing culture [119]. These preadapted cells, as well as those from other reports, either required serum supplementation for cloning [119, 134] or it was not specified that serum was not used for cell cloning approaches [135, 136]. Reports of serum-free cloning from biotech and pharmaceutical industry are rare, but examples include patent applications for the use of autologous feeder cells [137] or addition of supplements that support growth at cell densities below 100 cells·ml^{-1} [138]. Beside these approaches, gels such as agarose or high viscosity solutions like methylcellulose may be an alternative to retain cells as colonies or spheres under serum-free conditions [43]. The results presented here describe a method for serum-free outgrowth of cells deposited as single cells into microplate wells. Feeder cells, supplements, or even serum are not essential for cell outgrowth from 45% of the wells of a 96-well microplate. A statistical distribution of one cell per well results in approximately 35.5 wells not containing any cell [139]. Outgrowth of cells from 45% of all wells of a plate is thus equivalent to outgrowth from 71% of wells containing at least one cell. By using the serum-free single cell cloning strategy the risk of changes in product quality during time consuming adaptation phases of producer cells to serum-free growth as reviewed by Jenkins *et al.* [131] and regulatory concerns regarding the use of serum are avoided.

5 Synopsis

This section provides a summary of CEMAX host cell clones, which allowed generation of producer cells by site-directed integration of GOIs (Table 14). Additionally, several features of the expression system, which were predicted from theory are summarised below.

Highly similar proteins were expressed at nearly same rate after targeted integration of the GOI as shown with CEMAX host cell clone 01C090 and three isoforms of a potential immune suppressor fusion protein (Table 7).

The expression level of glycosylated fusion proteins was reduced by additional N-glycosylation sites and indicated limitations in the protein processing machinery. This was shown by the comparison of CRB-15, the potential immune suppressor protein, and the proprietary selectin fusion protein with host cell clone 01C090 and 25-004 (Table 7 and Table 10).

IgG were expressed at 1.7- to 2.5-fold higher rates than fusion proteins after integration of the GOI at the target site in host cell clone 25-004. This was not valid for host cell clone 01C090, which showed low recombination frequency and high variation in productivity indicating problems during recombination. The size of the replacement cassette could have been a reason for this.

Growth characteristics of host cells and producer cells derived thereof were highly similar. This was shown with CEMAX host cell clone 25-004 and producer clone 571034, which behaved highly similar regarding course of cell density, viability and cumulative volumetric cell days (Figure 35).

The predicted loss of the reporter and selection cassette used for selection of host cells after recombination was confirmed in case study II (section 4.3.3.3). No activity of the reporter hSEAP was detectable in cell culture supernatant of producer cells. Additionally, producer cells were sensitive to the selective agent hygromycin B indicating loss of the resistance and the reporter gene.

Table 14: Summary of CEMAX host cell clones and producer cell lines derived thereof.

XX: digits not common between clones numbers from one experiments; [*1]: low reproducibility when secreted glycoproteins were used as model protein; [*2]: host cell was assumed to be not clonal.

Host cell clone	Model protein	Producer clone	Specific productivity	Cross reference
07-022[*1]	GFP	Not specified	Not applicable	Figure 19, section 4.3.2.1
08-018[*1]	GFP	Not specified	Not applicable	Section 4.3.2.1
01C090	GFP	Not specified	Not applicable	Figure 19
01C090	Potential immune suppressor	Pool of producer clones	2.7 pg·c^{-1}·d^{-1}	Table 7, Figure 30
01C090	CRB-15	Several clones	1.9±0.8 pg·c^{-1}·d^{-1}	Table 7
01C090	Proprietary IgG	Several clones	1.6±0.8 pg·c^{-1}·d^{-1}	Table 7
25-004	Potential immune suppressor	22201	2.9±0.1 pg·c^{-1}·d^{-1}	Table 10, Figure 34
25-004	CRB-15	Several clones, 2720XX	2.2±0.1 pg·c^{-1}·d^{-1}	Table 10, Figure 37
25-004	Soluble selectin fusion protein	571034	2.0±0.1 pg·c^{-1}·d^{-1}	Table 10, Figure 35
25-004	Proprietary IgG	Several clones, 2710XX	4.1±1.3 pg·c^{-1}·d^{-1}	Table 10, Figure 43
25-004	ATROSAB	Not specified	5 pg·c^{-1}·d^{-1}	Table 10
15-013[*2]	Potential immune suppressor	202010	10.8±0.3 pg·c^{-1}·d^{-1}	Figure 31
15-013[*2]	Potential immune suppressor	202011	6.6±0.3 pg·c^{-1}·d^{-1}	Figure 31
15-013[*2]	Potential immune suppressor	202027	9.9±0.2 pg·c^{-1}·d^{-1}	Figure 31, Figure 33
31-26	CRB-15	Several clones	4.7±0.6 pg·c^{-1}·d^{-1}	End of section 4.3.3.1

6 Conclusion

Production of recombinant proteins in mammalian cells requires an intensive clone screening process to obtain stable high producer cells. To overcome this, a system for rapid generation of stable producer cell lines based on DNA double-strand break induced homologous recombination was developed in this work. The concept of the expression system was realised by generating modified host cell lines, which contained the tag vector at a transcriptional highly active site (recipient site). This genomic tag comprised cleavage sites for the homing endonuclease I-SceI. Targeted integration of a protein encoding gene was triggered by I-SceI mediated cleavage of the tag. Subsequent repair of the DNA lesion was catalysed by the cellular homologous recombination machinery with the replacement vector containing the product gene as a repair matrix. Upon this, the gene of interest got integrated at the previously tagged site and was expressed in a reproducible manner.

The system was based on CHO-K1 cells and allowed the use of serum-free and even chemically defined media through all steps of development. The basis for this was the 14-CHO-S cell line at the beginning of this work. A change in the source of a single medium compound made single cell cloning impossible and showed how sensitive single cells deposited into 96-wells are to media changes. This was overcome by screening of more than 20 commercial available media for their ability to support clonal growth. Of these media the chemically defined medium CDM3 showed superior growth characteristics combined with cloning efficiencies of approximately 15%. Cloning efficiency was further increased to 45% by temporary use of SFM12 for limited dilution cloning, which is equivalent to cell outgrowth from 71% of wells containing at least one cell. This high cloning efficiency was combined with viable cell densities of $1 \cdot 10^7$ c$\cdot$ml^{-1} and cumulative volumetric cell-days of $7.2 \cdot 10^7$ cell-days$\cdot$ml^{-1} demonstrated in batch experiments with 25-CHO-S cells in CDM3. In contrast to other reports of serum-free adapted CHO host cells [119, 134-138] no addition of serum, feeder cells or other supplements was necessary to promote single cell outgrowth. Thus this serum-free technology represents a good and regulatory compliant industrial standard for cell line development and process development.

Several predicted features of the expression system using site-directed integration could be verified (see also Synopsis). For example, the expression level of highly similar proteins after targeted integration was conserved. Three isoforms of a potential immune suppressor were produced at nearly identical levels. Differences in productivity were observed between antibodies and highly glycosylated fusion proteins. Productivities of 3 pg·c^{-1}·d^{-1} for fusion proteins and 6 pg·c^{-1}·d^{-1} for antibodies were achieved with host cell clone 25-004. The highest productivity for a glycosylated fusion protein achieved from a single gene copy after targeted integration was 10 pg·c^{-1}·d^{-1}. Compared to stable bulk pools of transfectants productivities achieved by gene targeting were 6.7-fold to 25-fold higher. These rates are sufficient for glycoproteins and early development stages including clinical phases I and II, but are too low for manufacturing of high volume products like IgG. A 15-fold higher product concentration was attained from the developed technology with glycoproteins compared to transiently transfected CHO cells expressing IgG [48, 49]. This was exemplified with a non-optimised fed-batch experiment in chemically defined medium and a product concentration of 150 mg·l^{-1} for a glycosylated fusion protein. The use of stable producer cells allows the scale-up to increased bioreactor volumes, which is difficult with the transient expression approach. Additionally it has been shown that growth characteristics of host cell clones and producer cell lines derived thereof were highly similar regarding viable cell densities, batch duration, viability and cumulative volumetric cell-days. This was demonstrated with 25-004 and producer clone 571034 and implied the possibility of a process platform with a pre-optimised cultivation strategy. Frequencies of targeted integration between 10^{-5} and 10^{-6} events per treated cell with a serum-free selection protocol and a total homology of 1598 bp were comparable regarding frequency and better regarding homology length compared to published results [61, 73, 105]. Gene targeting was confirmed at the molecular level with a PCR assay and by Southern blot analysis.

The expression system developed in this work is a regulatory friendly platform technology for protein expression from stable cell lines. It allows reproducible results regarding expression level and growth characteristics. Possibilities for implementation in basic and applied research include tests on secretion augmenting proteins, apoptosis inhibitors, proteins with metabolic activity, new promoters and other regulatory DNA elements. Furthermore the system allows the fast generation of stable producer cells for diagnostic proteins and structural analysis with low effort, which is economical regarding time, work and money. In drug development it saves time compared to transient expression and subsequent generation of stable production cells. Screening, fast production, and further development of drug candidates is feasible with one producer cell clone and shortens development time by approximately six months. A way of implementing this technology in the field of preclinical development was exemplified in case study I. Targeted integration and thus fast cell line development allowed the production of several protein isoforms from stable producer cells for an evaluation study. Once the decision towards a certain protein variant was driven, the producer cells from the initial experiment were recovered from a frozen cell bank and cultivated in larger scale for delivery of protein material for additional studies. Specific productivities of up to 10 $pg \cdot c^{-1} \cdot d^{-1}$ were sufficient for glycoproteins, but leave room for improvement, especially when IgG are produced. If this can be optimised in future, development of production cell lines would be the most profitable field of application.

The technology developed in this work is marketed as CEMAX® Technology for cell line development in 4 weeks (Celonic) and cGPS® CHO-S CEMAX® for research and *in vitro* screening purposes (Cellectis BioResearch).

7 Perspective

Although progress has been made, two major challenges remain. Resistant non-producer cells arose through selection of targeted cells by activation of neo via illegitimate recombination and zeo through promoter trap events. On the other hand, productivity was sufficient for glycoproteins and early development stages, but was too low for manufacturing of antibodies. These two points might be addressed using a modified selection system, transcription augmenting genetic elements, or as an alternative to CHO cells another host cell type.

Owing to background resistance a vector system with two non-functional portions of positive selection markers would be the best choice. These markers flank the exchangeable segment in the tag vector with the second marker being analogous to the Δneo used in this work. Activation of either resistance gene is impossible when the reporter cassette is removed after cleavage with I-SceI since no promoter remains. Promoter trap events from random integration of the replacement vector would be impossible as well, because the marker fragments corresponding to the genomic tag are non-functional. In addition, this vector design would allow higher flexibility in terms of the replacement cassette including free promoter choice and insertion of additional expression cassettes.

Productivity of CHO cells increases at higher copy numbers and seems to be low from single copy transgens [140]. This might be overcome with genetic elements such as insulators, S/MARs, UCOEs, and EASE that have shown to increase productivity of stable cells [29-35]. These elements might be a tool to support endogenous properties of the tag locus when implemented in the tag vector design. A combination of effects from these elements with the position effect at the integration site might facilitate delivery of higher productivity from a single copy.

Another chance to increase productivity of targeted cells to levels suitable for manufacturing could be the use of a host cell that is capable to deliver high productivities from low copy numbers. These could be the Per.C6® cell line, which has shown high level protein expression with relatively low transgene copy numbers [4] or the NS0 cell line. Hartman *et al.* demonstrated cell specific productivities between 20 and 58 pg·c^{-1}·d^{-1} with 4 antibody molecules expressed in a cholesterol independent NS0 cell line from a single transgene copy [130]. Although cell lines from these two examples have known limitations regarding their glycosylation characteristics, this might be overcome. Methods have been established to eliminate glycosyltransferases that cause deleterious sugar epitopes [141] and to introduce new enzymatic activity that confers positive effects on bioactivity [142, 143].

Isolation of host cells that allow generation of high producer cell lines by gene targeting may also be achieved by applying a higher throughput screening method like FACS combined with an additional selection step for recovery of clones that are suitable for homologous recombination. A higher throughput screening strategy and reasons for the necessity of such a selection step is described in appendix 8.2. Such a strategy should include a homologous recombination step using another homing endonuclease followed by selection of high producer host cells. Cells suitable for gene targeting that express a reporter gene with highest expression rate after the first gene replacement step could then be selected and used as host cells for generation of producer cells.

8 Appendices

8.1 Theoretical considerations on limited dilution cloning

A thought experiment was performed to address the question if it is beneficial to perform limited dilution cloning experiments as fast as possible after transfection and to determine the effect of cloning at particular times on the probability to find a particular clone.

Therefore, an imaginary situation with 1000 individual stable clones 24 h after transfection was created. Each clone carries a selection marker and has different characteristics like cell size, growth rate and productivity for instance. These stable clones are selected in a bulk pool by antibiotic selection to eliminate untransfected cells. Assuming that the clones do not differ in growth rate, each clone has probably proliferated to 1000 cells in a bulk pool containing 10^6 cells after 10 to 14 days of selection. Two individual limited dilution experiments are started 24 h and 10 to 14 days after transfection allowing the assessment of 1000 clones each time.

The cloning experiment started 24 h after transfection is a combination of selection of stable cells with dilution cloning. Clones derived thereof are referred to as mini pools. In this experimental set-up all 1000 individual clones are analysed, which is equivalent with a 100% probability that clone X is among them. When analysing 1000 clones in the second cloning experiment that was started when 1000 copies of each clone were available, the situation changes. It is now possible that several copies of particular clones are analysed and thus the total number of individual clones decreases. This situation can be described and calculated with a hypergeometric distribution (Formula 1). In this example the probability that clone X is among the analysed clones is reduced to 63% when the clones have proliferated to 1000 copies each. To reach a 95% probability that clone X can be found in this experimental set-up 3000 clones have to be analysed. For a probability of 99% this number increases to 4600 clones.

This basic thought experiment considers the selection of stable clones, especial the selection of a bulk pool, as a black box. The influence of the time of plasmid integration into the genome, the growth rates of individual clones or clone stability, for example, were not included in the calculation. For sake of demonstration the number of individual clones per transfection was set to 1000, what is at least a 10-fold underestimation in most experiments. For example, Wurm *et al.* calculated 10000 individual clones per transfection [144]. Taking into account a frequency of 1.9% for stable clones generated with the Nucleofector™ technology (Table 1) the number of clones will increase to approximately 40000 after transfection of $2 \cdot 10^6$ cells. Assuming one good clone per 1000 clones and the analysis of 1000 clones from these larger pools reduces the positive effect of selection in mini pools to levels comparable to the bulk pool approach. The probability to find one of ten good clones is 65% after 24h and 63% after 2 weeks assuming 10000 individual clones, for example.

Hence, the smaller the population of stably transfected clones, or the larger the number of analysed clones is, the more benefit is received from early dilution cloning regarding the probability to find a particular clone. Additionally, early cloning allows the recovery of clones with slightly reduced growth rates with higher probability than from a bulk pool. If two clones with population doubling times of 21 h (clone A) and 27 h (clone B) are selected in a bulk pool for 14 days, clone A has undergone 16 cell divisions compared to 12.4 doublings of clone B. Clone A would be represented as a population that is 11.8-fold larger than that of clone B with more than 65000 copies versus 5600 copies of clone B. One can assume easily that clone B is analysed at lower probability compared to clone A when cloning is performed from a bulk pool of stable transfectants.

On the other hand cloning from bulk pools of stable clones allows triggering towards recovery of clones with faster growth under conditions chosen during selection.

8.2 Host cell line generation 3 and 4: higher throughput screening

A higher throughput screening method based on fluorescence-activated cell sorting (FACS) was used as an alternative to the conventional clone screening process via limited dilution cloning. CEMAX host cells with high productivity should have been selected out of a larger number of transfectants compared to the clone screening by limited dilution. These candidates were selected for high level of intracellular reporter protein GFP by flow cytometry. Afterwards these cells were characterised for their protein secretion rate with the alkaline phosphatase, which was already used in development of the first and second host cell line series. Although productivity of the secreted reporter hSEAP was increased by the higher throughput screening, clones of host cell line generations 3 and 4 were not feasible for gene targeting. It was not possible to recover producer cells after selection of cotransfected cells.

Prior start of the experiments the tag vector was modified by introducing a fluorescence marker for sorting. The gene for the GFP was inserted downstream the hSEAP reporter into the tag vector as illustrated in Figure 48. Translation of the fluorescent protein was initiated at an IRES within the mRNA containing the open reading frames of both reporter genes.

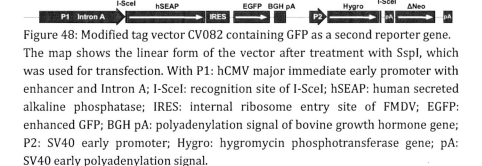

Figure 48: Modified tag vector CV082 containing GFP as a second reporter gene. The map shows the linear form of the vector after treatment with SspI, which was used for transfection. With P1: hCMV major immediate early promoter with enhancer and Intron A; I-SceI: recognition site of I-SceI; hSEAP: human secreted alkaline phosphatase; IRES: internal ribosome entry site of FMDV; EGFP: enhanced GFP; BGH pA: polyadenylation signal of bovine growth hormone gene; P2: SV40 early promoter; Hygro: hygromycin phosphotransferase gene; pA: SV40 early polyadenylation signal.

Cell line generation 3 was selected in a first approach, which was performed on a FACS Vantage SE at the institute for technical chemistry at Leibniz University Hannover. The result of this experiment is outlined in Figure 49. Mean fluorescence was stepwise increased in two rounds of bulk sorting out of an initial population of $8 \cdot 10^7$ stable 14-CHO-S/CV082 cells. The cells were transfected with 0.02 pmol tag vector CV082 and selected as a bulk pool. In these two rounds of sorting the initial cell population was reduced to 45000 cells due to repeated elimination of negative and low producer cells and low viability after sorting. A third round of both bulk and single cell sorting failed since it was not possible to recover the cells. Mean fluorescence was nearly doubled from the first sorting round to the second round with cells showing already homogeneous fluorescence (Figure 49 B).

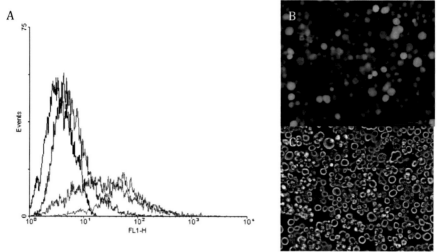

Figure 49: Increase in GFP fluorescence through repeated rounds of sorting.
A: overlay histogram with negative cells (black), bulk pool cells (blue), enriched cells after the first round of sorting (green) and after the second round of sorting (red). B: GFP positive cells after two rounds of sorting UV light with GFP filter setting and visible light (C). The bar indicates 100 µm.

A comparison of productivities of the secreted reporter protein hSEAP is shown in Figure 50. The initial bulk pool for sorting, which was transfected with 0.02 pmol DNA, produced levels of phosphatase activity less than the detection limit of the assay. This indicated a small proportion of producer cells or extreme low specific productivity. The productivity of a stable bulk pool of cells that were transfected with a regular amount of DNA was about 8 nU phosphatase activity per cell and day, for example. Nevertheless the stable bulk pool transfected with the low DNA amount was used for sorting, since GFP positive cells were detectable with the cytometer. After the first round of bulk sorting productivity was already approximately 10-fold higher than that of the bulk pool used for comparison. The population after two rounds of bulk sorting exhibited a productivity that was comparable to that of generation 1 host cell clone 01C090. The productivity data for the secreted protein were consistent with the intracellular level of GFP detected with the cytometer (Figure 49 A).

114

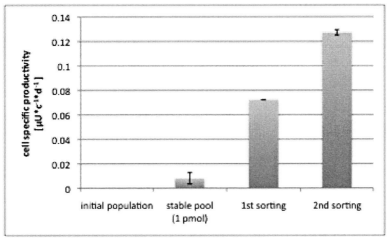

Figure 50: Productivity after bulk pool sorting.

The reporter protein hSEAP was not detectable in the initial population, which was a bulk pool of stable cells transfected with 0.02 pmol tag vector. In contrast to the pool used for sorting the bulk pool transfected with a regular amount of DNA (1 pmol) produces detectable levels of the reporter protein. The error bar of this population indicates deviation in productivity during cultivation in T-flasks while error bars of populations after first and second sorting round indicate deviation between duplicates of a 6-well assay.

Clonal cell lines were generated by limited dilution cloning in BD Select CHO medium based on the cell population from the second round of sorting. Unfortunately single cell sorting with the FACS was not successful owing to low cloning efficiencies. Albeit cloning efficiency in this medium was below 0.5%, 216 clones were analysed for expression of hSEAP. Most of these clones were recovered by using untransfected cells as feeder cells that supported clonal growth. As exemplified in Figure 51, up to 30% of clones isolated from the population that was enriched for high producer clones were identified as outperformer clones. This is in contrast to previous experiments where it was not possible to isolate high producer clones or even outperformer clones from bulk pools transfected with only 0.02 pmol tag vector and illustrated the power of the FACS technology to enrich high producer cells.

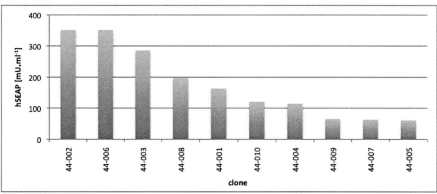

Figure 51: Analysis of hSEAP activity in 96-well supernatant of cloning experiment LD44.

Three outperformer clones were obtained in this experiment.

The best out of 216 clones were analysed for their productivity in more detail and compared with clones of first two host cell line generations. Figure 52 shows the analysis and comparison of selected clones.

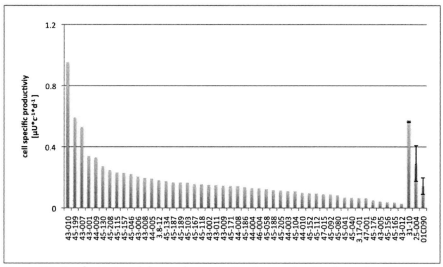

Figure 52: Productivity of clones derived after 2 rounds of bulk pool sorting.

Phosphatase activity was measured from supernatant of a 3-day 6-well assay. Productivity of clones from generation 1 and 2 was taken from Figure 18 and Figure 25.

The best clone generated with the FACS strategy, 43-010, exhibited a 1.7-fold higher productivity compared to the best clone generated with the traditional strategy. Clone 45-199 and 43-007 produced SEAP at a similar rate compared to clone 31-10. The other clones were in the range of 25-004 or below.

Unfortunately the high productivity was not conserved after targeted integration of the CRB-15 gene. Contrary to the expectation of higher productivity, no producer cells could be identified from 17 host cell clones tested. The only exception was achieved with clone 43-004, whose derivates produced CRB-15 at a rate of 0.2 $pg{\cdot}c^{-1}{\cdot}d^{-1}$.

In parallel to gene targeting experiments genomic DNA of selected clones was submitted to Southern blot analysis at a cooperation partner. The result of the analysis, which is depicted in Figure 53, gave in part an explanation why gene targeting was not successful.

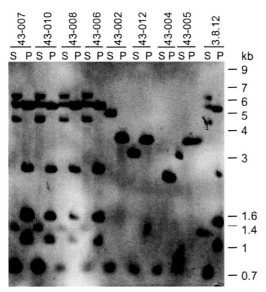

Figure 53: Southern blot analysis of clones from FACS enriched pools.
Restriction fragments containing sequences of the CMV promoter were detected with a compatible probe. The minimal possible size of fragments calculated from plasmid sequence were PvuII (P): 1.8 kb; SacI (S) 1.0 kb plus an internal fragment of 0.7 kb. Fragment size from a DNA molecular weight standard is shown on the right of the blot.

Most clones contained several copies of the tag vector what might cause lethal genome translocations after introduction of I-SceI activity during targeting, if these are located on different chromosomes [89]. In addition to high copy numbers, only four different integration patterns from nine clones were detected. This indicated a low heterogeneity in the sorted population. For example, clones 43-007, 43-010, 43-008, 43-006 and 3.8-12 showed an identical integration pattern according to Southern blot analysis. This was not expected after analysis of productivity with 5-fold differences between clone 43-010 and 3.8-12. One possible reason for this might have been that these clones were derived from one parental clone after integration of the tag vector. A phenotypic drift in subsequent cultivation including limited dilution cloning might be the reason for the heterogeneity [145]. Another possibility, albeit unreasonable with the background of the high number of identical fragments between the clones, might be transgene integration into highly repetitive DNA. The sequence directly surrounding the integration site would be identical resulting in similar fragment sizes, but the genomic context would be different.

Clone 43-004, the only host cell clone which allowed recovery of recombinant cells from gene targeting experiments, did not show a detectable fragment from SacI treated DNA except for the internal control fragment (0.7 kb). This indicated an incomplete integration of the promoter region and was a possible reason for low productivity after targeted integration of the CRB-15 gene. The remaining clones with intact single copy integrations were clone 43-002, 43-012 and 43-005. The data obtained with the probe for detection of CMV sequences were consistent with the Southern blot analysis for detection of sequences from the neo marker (data not shown). These clones did not allow the recovery of producer clones in several cotransfection experiments.

A second approach was performed with cells transfected with either 1 pmol tag vector, or 0.02 pmol using a different culture medium and a larger population of transfected cells. GFP positive cells were selected in 5 iterative rounds of sorting of cells cultivated in chemically defined medium CDM3. In total 10^9 cells, which was 10 times more than in the first approach, were sorted in the first round to achieve a higher degree of heterogeneity in the sorted pools. Limited dilution cloning experiments were performed after each round of sorting to avoid problems with low complexity of the sorted pool. In total 4334 clones were evaluated for productivity of the reporter protein hSEAP. Of these 30 clones and 5 sorted populations were tested in gene targeting experiments, but no producer cells could be recovered in these experiments. Southern blot analysis of these clones revealed a mixture of multiple integrations and a fraction of clones with incomplete integration of the tag vector (data not shown).

Although host cell clones from these higher throughput screening approaches did not allow recovery of high producer cells by targeted integration, it was shown that application of FACS is suitable for generation of high producer clones. This was consistent with published results [146-148]. The screening strategy was applied to select high producer cells from pools of cells transfected with low quantities of DNA, which was not possible by limited dilution cloning in previous experiments.

Nevertheless, the FACS approach may be applied successful under modified conditions in future. Host cells suitable for gene targeting experiments might be obtained, if a step focused on recovery of cells that allow gene replacement is included in the selection strategy. The host cell selection itself would have to include a homologous recombination step using a second homing endonuclease. Cells suitable for gene targeting that express a reporter gene with highest expression rate after the first gene replacement step are then selected and used as host cells for generation of producer cells.

8.3 Materials

Chemicals were obtained from Applichem, Fluka, Sigma, Merck and Roth if not specified else wise. General consumables like pipette tips were from Rainin, Peqlab, and Molecular BioProducts. Pipettes and centrifuge tubes were from Corning, reaction tubes from Sarstedt, sterile filters from Millipore and Pall. Photometer cuvettes were obtained from Roth, syringes from B.Braun.

8.3.1 Materials for molecular biology

8.3.1.1 Chemical competent cells

Transformation competent E.coli cells (Table 15) were used depending on the application. *E.coli* INV110 was used for preparation of unmethylated plasmid DNA and not for routine transformations.

Table 15: *E.coli* strains for transformation and their genotype.

Strain	Supplier	Genotype
INV110	Invitrogen C7171-03	*F′ [tra36 proAB lacI lacZM15] rpsL (Str) thr leu endA thi-1 lacY galK galT ara tonA tsx dam dcm supE44 Δ(lac-proAB) Δ(mcrC-mrr)102::Tn10 (TetR)*
NEB 5-alpha	NEB C2987H	*fhuA2Δ(argF-lacZ)U169 phoA glnV44 Φ80 Δ(lacZ)M15 gyrA96 recA1 relA1 endA1 thi-1 hsdR17*
XL1Blue	Stratagene 200130	*recA1 endA1 gyrA96 thi-1 hsdR17 supE44 relA1 lac [F′ proAB lacIZΔM15 Tn10 (Tet)]*

8.3.1.2   Enzymes

Table 16: Enzymes and suppliers for molecular biology.

Enzyme	Supplier
Restriction endonucleases	New England Biolabs, Invitrogen, Fermentas
Antarctic Phosphatase	New England Biolabs
T4 DNA Ligase	New England Biolabs
DNA Polymerase I, Large Fragment (Klenow)	Invitrogen
Taq DNA Polymerase	New England Biolabs
AccuPrime™ Pfx DNA Polymerase	Invitrogen

8.3.1.3   Synthetic oligonucleotides

Oligonucleotides for cloning applications were purchased from Eurogentec as OliGold oligonucleotides, which have a purity grade equivalent to that obtained by HPLC purification. PCR primers for confirmation of gene targeting were obtained from Microsynth as desalted oligonucleotides.

Table 17: Synthetic oligonucleotides.

Oligonucleotides 1-7 were obtained from Eurogentec, 8-11 from Microsynth.

No.	Name	Sequence (5' → 3')	bp
1	I-SceI_sense	AGC TTA GGG ATA ACA GGG TAA T	22
2	I-SceI_antisense	AGC TAT TAC CCT GTT ATC CCT A	22
3	P5-Hygro-pA	AAA AGC CTG AAC TCA CCG CGA CGT C	25
4	P3-Hygro-pA_PciI	GCT GGC CTT TTG CTC ACA TGT TCT TTC C	28
5	P5-MN-FRT-Neo-pA	TAC TAG GGA TAA CAG GGT AAT CTG ATC AGC ACG TGT TGG	39
6	P3-MN-FRT-Neo_PciI	GCT AGC TAC ATG TAA TAT TCG AAG CAG CGT CGA CGG T	37
7	P5-Hygro-pAc	GCA GGC CTA GGC TTT TGC AAA GAT CGA TAT GAA AAA GCC TGA ACT CAC CGC GA	53
8	5P_CEMAX-PCR1	GGC GTG TAC GGT GGG AGG TCT ATT AAA GC	29
9	3P_CEMAX-PCR1	GTG GAG CCA AAC GCA GTA CAA AGT GTT ACC	30
10	5P_CEMAX-PCR3	CGC CTC TGC CTC TGA GCT ATT CCA	24
11	3P_CEMAX-PCR3	CGT CAA GAA GGC GAT AGA AGG CGA	24

8.3.1.4 Plasmid vectors

The following plasmid vectors were used in this thesis:

CV001 - This proprietary expression vector of Celonic was used as the source of the promoter, enhancer and 3' untranslated region of the immediate early gene of hCMV [13, 15], the cloning site, the polyadenylation signal of the bovine growth hormone gene [16], and the neo expression cassette [18]. The neo cassette is composed of the SV40 early promoter, the neomycin phosphotransferase gene, and the SV40 early polyadenylation signal.

CV001/hSEAP – This proprietary reporter construct (Celonic) served as source for hSEAP that was used in the construction of the tag vector.

pCMV/Neo/EGFP – This vector (Celonic, [120]) was used for excising the truncated neo marker for the replacement vector. Additionally the gene for EGFP, which exhibits an absorption maximum at 488 nm and an emission maximum at 507 nm, was used for tag vector generation 3.

pcDNA5/FRT – This commercial cloning vector (Invitrogen) was used as source for PCR amplification of the hygromycin phosphotransferase gene [19].

pFUSE-Fc – This commercially distributed vector (Invivogen) was used in construction of IgG replacement vectors. It contains CH2 and CH3 domains and the hinge region of human IgG1. The expression cassette that was transferred into the replacement vector contained a composite promoter composed of the hEF-1α [10] combined with 5' untranslated region derived of human T-cell leukaemia virus [127], the IL2 signal sequence, and SV40 late polyadenylation signal [17].

pMONO-zeo-GFP – This commercial available vector (Invivogen) served as a source for the IRES-Zeo cassette for the replacement vector and the GFP-IRES-Zeo cassette for cloning of the GFP control replacement vector. The IRES is derived of foot and mouth disease virus [149]. The vector contains the red shifted GFP variant LGFP with absorption maximum at 480 nm and emission maximum at 505 nm.

pCLS0197 – The I-SceI expression vector was kindly provided by Cellectis bioresearch. Expression of the homing endonuclease is controlled by the hCMV promoter and enhancer.

pmaxGFP – The GFP expression plasmid (Lonza) was used for determination of transfection efficiency.

8.3.2 Materials for cell culture
Consumables like shake flasks were obtained from Corning. Cryotubes, tissue culture flasks, dishes and plates were from TPP.

8.3.2.1 Chinese hamster ovary cell lines
CHO cells, 14-CHO-S, were derived from CHO-K1 (ATCC® CCL-61™). These cells were previously adapted to growth in serum free medium (BD CHO medium) by Zahn *et al.* [132]. The cells were frozen at passage 14 (primary seed bank) and 19 (working cell bank). Due to discontinuation of BD CHO medium during this thesis

the culture medium changed to BD Select CHO medium, which had the same formulation but contains peptone from a different source.

23-CHO-S cells were a derivate of 14-CHO-S cells, obtained by adaptation to chemical defined medium CDM3 after feasibility was shown in this work. Cells were frozen at passage 19. 25-CHO-S cells were derived from CHO-K1 (ATCC® CCL-61™) by direct adaptation to chemical defined medium CDM3. Cells were frozen at passage 14.

8.3.2.2   Media and supplements

Media that were used for cultivation of CHO cells are listed in Table 18. The test media for the medium screening approaches were commercial available catalogue products, except for CDM1, CDM2, and CDM3. These media were encoded according to the type of medium for confidentially reasons. CDM was used to name chemically defined media and SFM for serum-free media containing hydrolysates. Cell Boost 4 (Thermo Scientific) and DS 100 Soy peptone (BD Biosciences) were used as feed media. Other supplements including selective antibiotics are listed in Table 19.

Table 18: Media used for cultivation of CHO-S cells.

Medium	Supplier	Catalogue number
BD CHO Medium	BD Biosciences	220229
BD Select™ CHO Medium	BD Biosciences	220253
F-12K	Invitrogen	21127-022

Table 19: Cell culture supplements.

Supplement	Supplier	Catalogue number
human insulin, recombinant	Sigma	91077C
L-glutamine	Invitrogen	25030-024
D-(+)-Glucose	Sigma	G8769
G418	Invitrogen	11811-031
Hygromycin B	Invivogen	Ant-hm-5
Zeocin	Invitrogen	R250-01
Foetal Bovine Serum	Invitrogen	16000-044

8.3.3 Antibodies

Table 20: Antibodies used for analytics and Cytometry.

Antibody	Application	Supplier, catalogue number
F(ab')₂ Fragment Rabbit Anti-Human IgG	Capture antibody, ELISA	Jackson Immunoresearch, 309-006-008
Peroxidase AffiniPure F(ab')2 Fragment Goat Anti-Human IgG, Fcγ fragment specific	Detection antibody, ELISA	Jackson Immunoresearch 109-036-098
R-Phycoerythrin conj AffiniPure F(ab')₂ goat anti human IgG, Fcγ fragment specific	Cytometry	Jackson Immunoresearch 109-116-098

8.3.4 Buffers and solutions

2x SEAP buffer
- 2 M Diethanolamine
- 1 mM $MgCl_2$
- 20 mM L-Homoarginine
- pH 9.8

SEAP substrate buffer
- 120 mM *p*-nitrophenol phosphate
- in 1x SEAP buffer

Erythrosine B solution
- 0.45 mM Erythrosine B
- 13.9 mM NaCl
- 0.44 mM KH_2PO_4
- pH 7.25 with NaOH

LB-Miller
- 5 g·l⁻¹ Yeast extract
- 10 g·l⁻¹ Soy peptone
- 10 g·l⁻¹ NaCl
- autoclaved 20 minutes at 121°C

EDTA (0.5 M; pH 8.0)
- 0.5 M EDTA
- pH 8.0 with NaOH

8 Appendices

5x TBE for agarose gel electrophoresis	0.445 M	Tris
	0.445 M	Boric acid
	0.01 M	EDTA
	pH 8.3	

TE buffer	100 mM	Tris
	10 mM	EDTA
	pH 8.0	

| Sodium acetate 3 M; pH 5.2 | 3 M | Sodium acetate |
| | pH 5.2 with acetic acid | |

PBS Dulbecco's	137 mM	NaCl
	2.7 mM	KCl
	8.1 mM	Na_2HPO_4
	1.47 mM	KH_2PO_4
	pH 7.4	

| Blocking reagent (ELISA) | 3% | BSA fraction V |
| | in PBS Dulbecco's | |

| Washing reagent (ELISA) | 0.5% | Tween 20 |
| | in PBS Dulbecco's | |

Substrate buffer (ELISA)	0.1 M	Citric acid
	0.1 M	K_2HPO_4
	300 mg*l^{-1}	OPD
	400 μl*l^{-1}	H_2O_2 (30%)

| HCl, 3 M | 99.5 ml | HCl 25% |
| | Diluted in 250 ml H_2O | |

Dilution Medium (ELISA)   10%   FBS

                                           in culture medium

PBS + 0.5%BSA   0.5%   BSA fraction V

                                           in PBS Dulbecco's

8.4 Methods

Methods used in this work include DNA cloning techniques, cell culture methods, and analytical methods and are described below.

8.4.1 DNA cloning techniques

Molecular cloning of recombinant plasmid DNA was accomplished according to methods described by Sambrook and Russell [150]. Compatible DNA-fragments were generated by restriction, fragments of interest were purified and linked together using a DNA ligase. Recombinant plasmids were then transformed into chemical competent *E.coli* cells and purified by mini-scale plasmid preparation. The recombinant plasmid of interest was identified by its characteristic DNA-fragmentation pattern visualised by analytical agarose gel electrophoresis after control restriction.

8.4.1.1 Fragmentation of DNA

Restriction endonucleases cut DNA at specific restriction sites. Restriction experiments were designed according to the reaction conditions recommended by the supplier. For preparation of DNA-fragments 2 µg plasmid DNA were treated with 5 to 10 unit endonuclease for 2 h. Control restrictions were performed using of 1 µl plasmid DNA from mini-scale preparation with 1 to 2 units endonuclease. Experiments were carried out in a reaction volume of 20 µl or at least the ten-fold volume of endonuclease to ensure glycerol concentrations below 5%. Double digestions were performed either in a compatible buffer, or sequential. Enzymes were inactivated by heat according to the manufacturer's recommendations or by column purification (8.4.1.5) if necessary.

8.4.1.2 Ethanol precipitation of DNA

Unless performed by column purification (see section 8.4.1.5), recovery of nucleic acids from aqueous solutions by ethanol precipitation was used.

According to Sambrook and Russel [150], concentration of monovalent captions in a DNA solution was adjusted to 0.3 M with sodium acetate (3 M; pH 5.2). After addition of 2 volumes ethanol (-20°C) the solution was stored at -20°C for 15 minutes to allow formation of precipitates. The precipitate was then recovered by centrifugation (16000 g; 10 minutes; 4°C). After removal of supernatant 70% ethanol was added to the DNA precipitate followed by a centrifugation step as above. Supernatant was removed, the pellet dried at room temperature and subsequently dissolved in TE buffer or a buffer for further enzymatic reactions.

8.4.1.3 Modification of DNA fragments

Blunt ends were formed by removal of 3'-overhangs or fill-in of 5'-overhangs using DNA Polymerase I, Large Fragment (Klenow). Klenow is a proteolytic product of *E.coli* DNA Polymerase I that has lost 5'→3' exonuclease activity and thus does not degrade 5'-termini. While the complementary strands of 5'-overhangs are filled-in by 5'→3' polymerase activity, 3'-overhangs are removed due to the 3'→5' exonuclease activity. The reaction was performed as follows: 1 µg DNA from an inactivated or purified restriction reaction was treated with 2 unit Klenow in a 60 µl reaction mixture in 1x restriction enzyme buffer supplemented with 33 µM each dNTP. The reaction was initiated by incubation on ice for 20 minutes and then stopped by addition of ethylenediaminetetraacetic acid (EDTA; 0.5 M; pH 8.0) to a final concentration of 10 mM followed by heat inactivation at 75°C for 20 minutes.

Dephosphorylation, the removal of terminal 5' phosphate groups, was used to suppress recircularisation of vector fragments during ligation. However, a compatible insert DNA fragment will still be ligated to the dephosphorylated vector fragment and the background of recircularised plasmid after transformation can be reduced. In this work Antarctic Phosphatase was used for dephosphorylation according to the protocol provided by the manufacturer. Only vector fragments with blunt-ends and those vector fragments generated by restriction with a single endonuclease were treated with the phosphatase.

8.4.1.4 Agarose gel electrophoresis

Agarose gel electrophoresis was used to separate, identify and isolate DNA fragments of specific sizes. Gels were prepared with low electroendoosmosis (EEO) agarose and run in horizontal configuration in a constant electric field of 4 to 5 V·cm^{-1} in 1x TBE buffer. The gel strength was between 0.8% and 2.5%. Staining of DNA was carried out by fluorescent intercalating agent ethidiumbromide, which was added into the gel matrix prior electrophoresis at a concentration of 0.5 µg·ml^{-1}. Samples were supplemented with 5x Hi-density TBE sample buffer (Invitrogen) to 1x concentration prior application and 2-Log DNA ladder (New England Biolabs) served as a molecular weight standard.

8.4.1.5 DNA clean up

The NucleoSpin® Extract II Kit (Macherey-Nagel) was used for purification of DNA-fragments after excision from preparative agarose gels, enzymatic reactions, and PCR according to the recommended protocol. Recovery of DNA is based on silica-membrane technology.

8.4.1.6 DNA ligation

One of the applications of DNA ligase *in vitro* is their use as an enzyme for the formation of phosphodiester bonds between 3'-hydroxyl and 5'-phophoryl termini of DNA fragments. For instance, vector fragments can be linked to isolated DNA fragments to generate a recombinant plasmid during molecular cloning. T4 DNA Ligase was used in this work for joining vector and insert DNA fragments at a molar ratio of 1 to 3. Ligation of blunt-end fragments was performed using 100 ng vector DNA. The reaction mix was incubated for 2 hours at 16°C. For sticky-end ligations the vector amount was reduced to 50 ng and incubation was performed at room temperature for 10 minutes.

8.4.1.7 Transformation and cultivation of *E.coli*

Transformation chemical competent *E.coli* cells was performed according to the recommendations of the manufacturer with 5 µl of a ligation reaction or 10 ng plasmid DNA. Transformants were selected on LB-Miller (lysogeny broth) agar plates containing 50 µg·ml^{-1} ampicillin at 37°C. Further propagation for small scale plasmid DNA preparation was realised in 3 ml over night suspension cultures or 250 ml over night cultures for preparation of transfection grade DNA.

Suspension cultures were performed in LB-Miller supplemented with ampicillin to $100\ \mu g \cdot ml^{-1}$ at 37°C and agitated at 180 to 300 rpm. Zeocin was used as selective antibiotic at a concentration of $25\ \mu g \cdot ml^{-1}$.

8.4.1.8 Polymerase Chain Reaction

PCR is an *in vitro* method for amplification of DNA using a thermostable DNA polymerase [151]. DNA segments flanked by annealing sites for specific oligonucleotide primers are amplified at high temperature after initial template denaturation by recurring cycles of denaturation, primer annealing and DNA synthesis.

Primer design was performed using primer design software OLIGO [152]. It was considered not to include hairpin loops or other secondary structures, stretches of recurring nucleotides or sequences that tend to form homo- or heterodimers if possible. The complementary to the template extended over 18-24 nucleotides with a GC-content of 50% to 60%. Primers were designed to have a similar melting temperature which was not below 55°C. Restriction sites were incorporated into the primer on the 5'-end followed by up to 4 terminal nucleotides depending on the requirements of the restriction endonuclease. To achieve good binding characteristics a complementary G or C nucleotide was incorporated at the 3'-end of the oligonucleotide of primer.

8.4.1.8.1 Amplification of DNA fragments for subcloning

Exponential amplification of DNA segments for subcloning into plasmid vectors was performed using AccuPrime™ Pfx DNA Polymerase exhibiting 3'→5' exonuclease activity according to the recommendations of the supplier. The reaction mix contained 10 ng plasmid DNA template, 300 nM each primer, and 300 μM each dNTP in 1x reaction buffer containing $MgSO_4$. 1 unit polymerase was used in a 50 μl reaction. Restriction sites for subcloning were incorporated in the 5'-end of the oligonucleotide primer. DNA-fragments synthesised by PCR were purified with the NucleoSpin® Extract II Kit (section 8.4.1.5). Table 21 shows the PCR program for this application.

Table 21: PCR program for amplification of DNA fragments for subcloning.

Step	Description	Parameters
1	Initial denaturation	95°C, 4 min
2	Denaturation	95°C, 15 s
3	Primer annealing	melting temperature -4°C, 30 s
4	Extension	68°C, 1 min per kb DNA fragment
5	Repeated cycles of step	10 times cycle 2-4
6	Denaturation	95°C, 15 s
7	Primer annealing	melting temperature -4°C, 30 s
8	Extension	68°C, 1 min per kb DNA fragment + 5 s per cycle
9	Final extension	68°C, 10 min
10	Cooling	4°C

8.4.1.8.2 Colony PCR

Colony PCR was used for screening larger numbers of *E.coli* clones for a ligation product after cloning of a PCR fragment, for example. An *E.coli* lysate is directly applied in PCR to detect the plasmid clone of interest.

Colonies were picked with a pipette tip and transferred into 5 µl water in a PCR tube. The pipette tip was then used for inoculation of a 3 ml suspension culture for further analysis of positive clones. Cell lysis was achieved by incubation at 95°C. After 5 minutes the lysate was supplemented with 15 µl PCR mix (Table 22) and the reaction was initiated according to the PCR program in Table 23. Plasmid DNA containing an amplifiable DNA insert was used as a positive control. Therefore, 2.5 ng plasmid DNA instead of *E.coli* cells were used. Water was used as negative control.

The analysis of 8 µl PCR product from samples and controls was achieved by agarose gel electrophoresis (8.4.1.4). Clones enabling amplification of the PCR fragment of interest were further analysed by control restriction after small-scale preparation of plasmid DNA (8.4.1.9).

Table 22: PCR mix for colony PCR.

Final concentration in PCR is acquired after addition of 15 µl PCR mix to 5 µl cell lysate.

Component	Amount in PCR mix per reaction	Final concentration in PCR
ThermoPol Taq Reaction Buffer (10x)	2 µl	1x
Upstream primer (10 pmol·µl⁻¹)	1 µl	0.5 µM
Downstream primer (10 pmol·µl⁻¹)	1 µl	0.5 µM
dNTP-mix (10 mM each)	0.4 µl	0.2 mM
Taq DNA Polymerase (5 U·µl⁻¹)	0.2 µl	1 U per reaction
Water	10.4 µl	Final volume is 20 µl per reaction

Table 23: PCR program for Colony PCR.

First three cycles were performed at lower annealing temperature if length of complementary sequences to the template were reduced during cloning.

Step	Description	Parameters
1	Cell lysis	5 min, 95°C
2	Addition of PCR mix	95°C
3	Initial denaturation	3 min, 94°C
4	Denaturation	45 s, 94°C
5	Primer annealing	1 min, primer melting temperature -4°C
6	Extension	1 min/kb, 75°C
7	Repeated cycles of step 4-6	33 cycles
8	Final extension	5 min, 75°C

8.4.1.8.3 PCR-based confirmation of homologous recombination

Among other techniques, PCR was used to characterise clones after targeted integration of a gene of interest. Homologous recombination was confirmed by designing PCR primers that allow amplification of a DNA fragment only if primer binding sites are combined in the recombined integration site. No PCR product can be amplified from the genomic tag or the replacement vector alone.

Homologous recombination at 5'-end of the integration cassette was confirmed with 5'-primer that hybridises within a region of the CMV promoter located in the tag vector but not in the replacement vector. The corresponding 3'-primer hybridises with the IRES element of the replacement vector, which is part of the expression cassette after homologous recombination.

For the 3'-end of the integration cassette the sequence for the 5'-primer is derived from the SV40 early promoter in the replacement vector, which drives the expression of the neo marker after targeted integration. The corresponding 3'-primer binds to the 3'-region of the neo gene during PCR. This region of the selection marker is not included in the replacement vector.

200 ng genomic DNA was diluted in 5 µl water prior addition of 20 µl PCR mix (Table 24). Water (negative) and 1 ng appropriate plasmid DNA (positive) served as controls. The final concentration of the components was achieved after addition of 20 µl PCR mix to 5 µl template. The PCR program is shown in Table 25. The analysis of 10 µl PCR product from samples and controls was achieved by agarose gel electrophoresis (8.4.1.4).

Table 24: PCR mix for confirmation of homologous recombination.

Component	Amount in PCR mix per reaction	Final concentration in PCR
Taq Reaction Buffer (10x)	2.5 µl	1x
Upstream primer (10 pmol·µl⁻¹)	1 µl	0.4 µM
Downstream primer (10 pmol·µl⁻¹)	1 µl	0.4 µM
dNTP mix (10 mM each)	0.4 µl	0.2 mM
Taq DNA Polymerase (5 U·µl⁻¹)	0.2 µl	1 U per reaction
Water	14.9 µl	Final volume is 25 µl per reaction

Table 25: PCR program for confirmation of homologous recombination.

Step	Description	Parameters	
1	Initial denaturation	4 min, 95°C	
2	Denaturation	30 s, 95°C	
3	Primer annealing	30 s,	63.2°C
4	Extension	72°C	PCR1: 2 min PCR2: 1 min
5	Repeated cycles of step 4-6	31 cycles	
6	Final extension	12 min, 72°C	

8.4.1.9 DNA isolation

Plasmid DNA introduced into *E.coli* cells via transformation was isolated using commercial available preparation kits according to the protocol provided by the manufacturer. Briefly, plasmid DNA was isolated from *E.coli* cells by a modified alkaline lysis method similar to the one described by Birnboim and Doly [153] and recovered after binding to an anion-exchange resin (NucleoBond PC 500 EF Kit, Macherey-Nagel) or a silica membrane in case of small-scale preparations with the NucleoSpin Plasmid Kit (Macherey-Nagel). Cells were harvested from 250 ml respectively 2 ml over night cultures.

Transfection-grade, endotoxin-free plasmid DNA was analysed by spectrophotometry (8.4.1.10), control restriction and agarose gel electrophoresis (8.4.1.4) prior transfection of mammalian cells to ensure plasmid identity and integrity.

Genomic DNA of CHO cells was isolated from $2 \cdot 10^6$ cells using the Wizard Genomic DNA Purification kit (Promega) according to the supplied protocol including RNAse A treatment.

8.4.1.10   Spectrophotometry of DNA

Nucleic acids, which absorb significantly at a wavelength of 260 nm, were quantified by spectrophotometric measurement. Thereby an optical density (OD) at 260 nm of 1.0 corresponds to a concentration of double-stranded DNA of 50 ng$\cdot\mu$l^{-1}. The ratio of OD at 260 nm and 280 nm was used as an estimate of purity. DNA solutions were diluted with water to achieve OD values between 0.1 and 0.8.

8.4.1.11   Linearisation of plasmid DNA for transfection

To increase frequency of full length vector sequences during stable integration, plasmid DNA was fragmented with SspI prior transfection when generating stable CEMAX host cells. Besides, parts of bacterial regulatory sequences (pUCori and ampicillin resistance gene) were removed from the plasmid. Restriction was performed according to section 8.4.1.1 using 50 µg plasmid DNA and 25 unit restriction endonuclease in a reaction volume of 200 µl. Incubation time was increased to 8 h due to the reduced amount of enzyme. After separating the fragments by agarose gel electrophoresis (8.4.1.4) the vector fragment was excised from the gel matrix and purified as described in section 8.4.1.5 with the following modifications to the standard protocol: depending on the size of the gel fragment 3 to 4 NucleoSpin® columns were used to recover the DNA fragment. An additional wash step with 500 µl binding buffer was used to ensure removal of residual agarose after binding of the DNA to the silica membrane. Following elution with 100 µl TE buffer (pH 8.5) DNA was concentrated by ethanol precipitation (see 8.4.1.2), dissolved in 10 µl TE buffer and quantified by spectrophotometry.

8.4.1.12   Southern blot analysis

Southern blotting is a method to confirm a DNA sequence in a complex DNA sample and was named after its inventor Edwin Southern [154]. The method combines separation of fragmented DNA via agarose gel electrophoresis, transfer to a filter membrane and probe hybridisation followed by detection of the hybridised probe. When performed with enzyme-digested genomic DNA of transfected cell lines it allows drawing conclusions about the rough number of transgene copies and the integration pattern. The analysis was performed at a cooperation partner according to a standardised protocol. Briefly, genomic DNA of $2 \cdot 10^7$ frozen cells was isolated, 10 µg thereof were used endonuclease treatment, separated by agarose gel electrophoresis, transferred to a nylon membrane. Fragments were then detected with the DIG System (Roche) according to the supplier's recommendations. For restriction endonuclease treatment enzymes were chosen, which cut within the known sequence to achieve fragments with a known minimum size.

8.4.2 Cell culture methods

All cell culture work was performed in sterile environment under a laminar flow hood without media supplementation with anti bacterial or anti fungal antibiotics except hygromycin B, G418 and Zeocin for selection of transfected cells.

8.4.2.1 Cultivation of 14-CHO-S and 25-CHO-S cells

14-CHO-S were cultivated in serum-free medium BD CHO supplemented with 4 mM L-glutamine and 25-CHO-S cells in chemically defined medium CDM3 supplemented identical. Cells were maintained in humidified CO_2-incubators (37°C, 5% CO_2, 95% relative humidity) using tissue culture flasks and plates, or shake flasks agitated on an orbital shaker (Ovan, 3 cm movement, 100 rpm), or using headspace aeration in glass bowl spinner flasks (50-100 rpm; 5% CO_2, 21% O_2, gas flow 2-4 sl$\cdot$h^{-1}).

14-CHO-S cells were routinely subcultivated at cell densities between $2 \cdot 10^6$ cells$\cdot$ml^{-1} and $4 \cdot 10^6$ cells$\cdot$ml^{-1} to an inoculation density of $3 \cdot 10^5$ cells$\cdot$ml^{-1} by diluting cells in fresh medium.

Culture medium was completely exchanged during selection of stable cells or when transferring cells into another culture medium for instance. Therefore cells were harvested by centrifugation (200 g, 5 minutes), supernatant was removed and cells were resuspended in fresh medium.

Fed-batch experiments were performed either by addition of DS 100 Soy peptone (3 g$\cdot$l^{-1}) or Cell Boost 4 (3.5 g$\cdot$l^{-1}) on day 3 followed by supplementation every second day. Glucose at levels <3 g$\cdot$l^{-1} and glutamine at levels <1 mmol$\cdot$l^{-1} were fed separately.

14-CHO-S cells and recombinant derivates were cultivated temporary in F12-K medium supplemented with 10% foetal bovine serum (FBS) for selection of targeted integrants in some experiments.

8.4.2.2 Media screening

The method for screening commercial available CHO media was direct adaptation, cultivation in shake flasks and assessing performance by batch and cloning experiments.

14-CHO-S cells were transferred into test media by a complete medium exchange with an inoculation density of $3 \cdot 10^5$ cells·ml^{-1}. The adaptation was carried out in 125 ml shake flasks in a working volume of 30 ml with subculturing steps every three days. The applicability of the test media to promote growth from single cells was evaluated in a limited dilution experiment on day 11 of the adaptation process. This was done by inoculating two 96-well microplates with 1 cell per well in each medium that promoted growth without aggregation of cells. Plates were incubated for two weeks and then screened for colony formation by microscopy to calculate the cloning efficiency. Furthermore a batch experiment was started on day 12 of adaptation by complete medium exchange with an inoculation density as described above. The batch experiment was run in duplicates until viability reached values below 70%.

Calculating a score according to the decision matrix shown in Table 12 on page 97 was used for the selection of the best medium. Therefore the best medium for each parameter got the full parameter score whereas other media got a fraction of it depending on the relative performance.

In a second screening round media were analysed for their ability to allow single cell cloning of 25-CHO-S cells by temporary use of the test medium for the cloning experiment only. The cloning experiment was performed as above by simply diluting the cells in the test media prior inoculum of microplate wells.

All test media were supplemented to a final concentration of 4 mM L-glutamine. In the second experiment CDM10 was additionally supplemented with insulin at a concentration of 10 mg·l^{-1}.

8.4.2.3 Transfection

Transfections were made using the Nucleofector™ technology with an optimised protocol for 14-CHO-S cells using the Cell line Nucleofector Kit V (Lonza). Per transfection $2 \cdot 10^6$ to $1 \cdot 10^7$ cells from exponential growth phase ($1 \cdot 10^6$ to

$2 \cdot 10^6$ cells$\cdot$ml^{-1}) were resuspended in 100 µl Nucleofector® solution V and supplemented with plasmid DNA from endotoxin-free preparations. Transfection was then carried out with program U-024 and transfected cells were transferred into prewarmed cell culture medium at a concentration of $1 \cdot 10^6$ cells$\cdot$ml^{-1}. For stable transfections cells were at least incubated for one day prior applying selection pressure. The control plasmid pmaxGFP was used for determination of transfection efficiency 24 h post transfection by determining concentration of viable and fluorescent cells. Transfection efficiency was then calculated according to Formula 7.

8.4.2.3.1 Generation of CEMAX host cells

For generation of CEMAX host cell lines $2 \cdot 10^6$ cells were transfected with 1 pmol tag vector linearised with SspI. In some experiments the DNA amount was reduced up to 0.0002 pmol.

8.4.2.3.2 Generation of CEMAX producer cells

Per experiment $5 \cdot 10^6$ CEMAX host cells were cotransfected with 8 µg I-SceI expression vector pCLS0197 and 0.7 pmol replacement vector. Previously these conditions were optimised by variation of amounts for both plasmids. After transfection cells were transferred into 2 ml prewarmed medium.

8.4.2.4   Limited dilution cloning

Limited dilution cloning was used to generate clonal cell populations and mini pools of approximately one to five individual clones. A parental cell culture was diluted to the inoculum concentration required for the experimental question by serial dilution in culture medium. 96-well tissue culture plates were inoculated with 100 µl diluted cell suspension per well and incubated in a humidified CO_2 incubator. After one day one plate was randomly inspected by microscopy to ensure correct inoculation. To counteract depletion of unstable media components feeding with 100 µl fresh medium was performed once a week. The plates were screened for cell growth beginning two weeks after inoculation.

Selection of stable mini pools was achieved by combining selection of stably transfected cells with limited dilution cloning. Therefore cell cloning was performed one day after transfection with an increased cell number (i.e. 10-90 cells per well) in selective culture medium.

Clonal cell lines were established by limited dilution cloning of a polyclonal cell pool or mini pool with an inoculum of 1 cell per well. Single cell derived clones have been checked for outgrowth from a single cell by microscopy within one day after inoculation.

Selection of stable producer cells generated with the CEMAX system developed was initiated directly after transfection. $2 \cdot 10^6$ cells were diluted into 96-wells with an equivalent of 4000 cells per well for mini pool selection. After 3 days G418 selection was initiated by addition of 50 µl culture medium with 3-fold G418 concentration. Another 3 days later 50 µl culture medium with 4-fold Zeocin concentration was added to achieve double section. 100 µl fresh medium was added another week later. If more medium was added in cases of longer selection periods 200 µl supernatant were removed prior addition of 100 µl fresh medium. The development of the protocol is described in section 4.3.1.

8.4.2.5   Clone screening

Rating and selection of clones generated in limited dilution cloning experiments was performed in at least three steps: First the mass of non-producers was excluded from further experiments by taking the product titer in 96-wells of the cloning experiment into account. Only the best clones (<10% of all clones) were expanded through 24-well to 6-well scale. The product titer in this second stage was used to confirm the first measurement and to exclude clones with low productivity. Measuring cell specific productivity (see 8.4.2.6) represents the third step in clone selection. This was done after cells were expanded to 5 to 20 ml scale. Results from southern blot analysis were further taken into account to exclude clones with multiple integrations of transfected plasmid sequences.

8.4.2.6   Productivity test

Cell specific productivity was calculated after cultivating a defined cell number in 6-well scale for three days and correlating cumulative volumetric cell-days and product concentration as described by Dutton et al. [155] according to Formula 5. Therefore $6 \cdot 10^5$ viable cells were transferred into 2 ml fresh culture medium in a

6-well by complete medium exchange. After three days cell density was determined and cell culture supernatant was harvested for quantification of product concentration. Alternatively productivity was determined during batch experiments and routine subcultivation or cell expansion using the same formula.

8.4.2.7   Antibiotic selection

Optimal concentration of selective antibiotics was determined by cultivating untransfected cells in a set of medium with different concentrations of an antibiotic (Table 26) and observing cell density and viability. Cells were inoculated at a concentration of $5*10^5$ cells$*$ml^{-1} by complete medium exchange into the respective medium. Medium was exchanged again after four days. Progression of cell death was monitored by determination of viable cell counts and viability every day. The minimum concentration for each antibiotic was chosen that kills cells to a viability of <20% within 6 days.

Table 26: Working concentration for antibiotic selection of stable cell lines. These concentrations were used, if not stated otherwise.

Antibiotic	Concentration
Hygromycin B	200 µg$*$ml^{-1}
G418 (Geneticin sulphate)	400 µg$*$ml^{-1}
Zeocin	200 µg$*$ml^{-1}

8.4.2.8 Thawing of cells

Thawing of cells was performed according to a method suitable for cells that are refractory to recovery from a frozen stock. Briefly, cells were thawed in a water bath (37°C) and then transferred into a 50 ml centrifuge tube for addition of increasing amounts of culture medium (4°C) as specified in Table 27. After centrifugation (200 g, 5 minutes) the cell sediment was resuspended in prewarmed culture medium and the cell density was adjusted to $3*10^5$ cells$*$ml^{-1}.

Table 27: Scheme for addition of culture medium to cell suspension during thawing.

The procedure is calculated for 1 ml of frozen cells.

Minute	Medium addition
0	0.1 ml
1	0.12 ml
2	0.15 ml
3	0.19 ml
4	0.26 ml
5	0.36 ml
6	0.52 ml
7	0.86 ml
8	1.69 ml
9	4.75 ml

8.4.2.9 Preparation of conditioned medium

Conditioned medium was obtained from cell cultures in the exponential growth phase by two serial centrifugation steps. Cells were removed from cell culture by centrifugation at $200\,g$ for 5 minutes and further precipitable material was eliminated from the supernatant in the second centrifugation step ($3000\,g$, 5 minutes).

8.4.2.10 Cryopreservation

$1{\cdot}10^7$ cells were cryopreserved in freezing medium composed of a mixture of 50% fresh and 50% conditioned medium supplemented with 7.5% DMSO. Therefore cells were transferred to 50 ml centrifuge tubes and centrifuged at $200\,g$ for 5 minutes at room temperature. The cells were resuspended at a cell density of $1{\cdot}10^7$ cells per millilitre in freezing medium, aliquoted into cryotubes and transferred into Nalgene cryopreservation boxes to allow a cooling rate of 1°C per minute. These boxes were stored at –80°C for at least 24 h before the vials were transferred into the vapour-phase of a liquid nitrogen tank or a deep freezer (-150°C).

8.4.2.11   Determination of clone stability

To determine stability of productivity of CEMAX clones producing a protein of interest these cells were cultivated in the absence of selection pressure over a period of approximately 60 cell generations. During the experiment cells were maintained in 125 ml shake flasks and subcultured to $2 \cdot 10^5$ cells$\cdot$ml^{-1} every 3 or 4 days. Compared to a reference culture cultivated under selection pressure the read out for stability were product concentration in batch and cell specific productivity combined with results obtained by cytometric analysis of cells with the Guava PCA after intracellular staining of the protein of interest. Batch experiments were performed in duplicates in non-selective medium with a seeding cell density of $5 \cdot 10^5$ cells$\cdot$ml^{-1} after complete medium exchange.

8.4.2.12   Flow cytometry

Flow cytometry allows the analysis of cellular characteristics like size, granularity, and relative fluorescence intensity at the single cell level. It was used as a higher throughput method to isolate high producer host cell clones and to analyse intracellular protein of interest for cell line stability studies.

8.4.2.12.1   Fluorescence activated cell sorting

FACS experiments as higher throughput screening tool were used for generation of host cell lone generation 3 and 4 at the Institute for technical chemistry at Leibniz University Hannover (FACS Vantage SE) and the Institute of Cell Biology at University Duisburg-Essen (FACS Diva), respectively. The cells were analysed for fluorescence of the intracellular reporter protein GFP. A stable transfected bulk pool was enriched for GFP positive cells in a first sorting round in enrich mode and in additional rounds in normal mode for higher purity. Minimum fluorescence level of gated cells was increased in each sorting round. After sorting cells were recovered in conditioned medium and were cultivated for at least 3 days prior the next sorting experiment. As sheath fluid either BD FACSFlow (BD Biosciences), or PBS (Biochrom) were used. BD Accudrop Fluorescent Beads and BD Calibrite (BD Biosciences) were used for the equipment set-up.

8.4.2.12.2 Analysis of intracellular protein of interest

Cytometric analysis of intracellular protein level was performed with the GUAVA NEXIN application after fixation of cells from culture and immunofluorescent staining of intracellular proteins containing the Fc region of human IgG.

Cell fixation was performed after harvesting $3 \cdot 10^6$ cells by centrifugation (300 g, 5 minutes), washing the cell pellet with 10 ml PBS followed by another centrifugation step and resuspension of the cell pellet in 3 ml 4°C cold 70% Ethanol. Fixed cells were stored at -20°C until analysis.

For sample preparation $2 \cdot 10^5$ fixed cells were centrifuged as above and washed in 10 ml PBS +0.5% BSA. After centrifugation and removal of supernatant the intracellular protein of interest was labelled with a PE conjugated antibody fragment specifically binding human IgG. Therefore cells were resuspended in 100 µl antibody solution diluted 1 to 20 in PBS +0.5% BSA and incubated for 30 minutes at room temperature in the dark. For analysis cells were resuspended in 300 µl PBS +0.5% BSA after centrifugation and removal of supernatant.

The Guava PCA was operated in the Guava NEXIN application according to the recommendations of the supplier. 25-CHO-S cells were used as negative control and were set to a relative fluorescence between 10^1 and 10^2. Data were exported in FCS2.0 format for preparation of overlay histograms employing WinMDI 2.8.

8.4.2.13 Spent media analysis

Analysis of cell culture supernatant for concentration of key nutrients like glucose and glutamine and by-products like lactate was performed using either the Ebio compact glucose analyser and the SYI 2700 select, or the YSI 7100 MBS. Measurement is based on enzyme electrode biosensors and was performed according to the recommendations of the supplier with cell free culture samples. Therefore cells were removed by centrifugation (400 g, 5 minutes).

8.4.2.14 Cell counting

Determination of cell concentration was performed manually by the erythrosine B dye exclusion method using a haemocytometer or by automated trypan blue exclusion method with a VICELL-XR (Beckman Coulter). The parameter settings for automated analysis are summarised in Table 28.

Table 28: Parameter settings for automated cell counting.

Parameter	Setting
Minimum diameter	15 μm
Maximum diameter	50μm
Cell brightness	87%
Cell sharpness	85
Viable cell spot brightness	93%
Viable cell spot area	13.5%
Minimum circularity	0.2
Decluster degree	medium

8.4.3 Analytical methods

This section describes analytical methods for quantification of protein products in cell culture supernatant via hSEAP assay and ELISA. Purified protein was processed by isoelectric focussing and BCA.

8.4.3.1 Quantification of secreted alkaline phosphatase

The colorimetric assay for quantification of hSEAP in cell culture supernatant was adapted from a method reported by Cullen [156]. To inactivate contaminating phosphatases samples or dilutions thereof were incubated at 65°C for 30 minutes in 1.5 ml reaction tubes and supplemented with an equivalent volume of 2x SEAP buffer. 100 μl of this solution was added to wells of a flat-bottomed 96-well microplate (Roth) in duplicates. The plate was incubated at 37°C for 10 minutes prior addition of 10 μl prewarmed SEAP substrate buffer per well. Plates were incubated at ambient temperature for 10 minutes, mixed, and then analysed using a microplate spectrophotometer by measuring the light absorbance at 405 nm. Placental alkaline phosphatase (Sigma) served as a standard and was assayed in serial two-fold dilutions ranging from 200 mU·ml⁻¹ to 6.25 mU·ml⁻¹. Culture medium was used to dilute samples and setting the blind value.

Analysis was streamlined for clone screening approaches by directly applicating 50 µl cell culture supernatant to the assay plate while samples, standards and blank were analysed in single measurement and were processed as described above.

Raw data were analysed using a linear regression with either SoftMax Pro, or Tecan software and exported to Excel for further interpretation.

8.4.3.2 Enzyme-linked immunosorbent assay

Human IgG and fusion proteins containing the Fc region of human IgG were quantified in cell culture supernatant using a sandwich ELISA as described in Table 29.

Table 29: Protocol for Fc γ specific ELISA.
The assay was used for quantification of human IgG and fusion proteins.

Component	Volume per well	Incubation time	Incubation temperature
Capture antibody, 5 µg·ml⁻¹ in PBS	100 µl	over night	4°C
PBS	200 µl	wash, 3 times	
Blocking reagent	200 µl	1 h	37°C
PBS	200 µl	wash, 1 time	
Samples	100 µl	1 h	room temperature
Washing reagent	200 µl	wash, 4 times	
PBS	200 µl	wash, 1 time	
Peroxidase conjugated detection antibody, diluted 1/5000 in PBS	100 µl	45 min	room temperature
Washing reagent	200 µl	wash, 4 times	
PBS	200 µl	wash, 1 time	
Substrate buffer	100 µl	10 min	room temperature
HCl, 3 M	50 µl	1 - 60 min	room temperature

Briefly, Fc γ specific capture antibody was coated to a flat bottom 96-well plate (NUNC MaxiSorb™). To avoid unspecific reactions, non-specific binding sites were blocked with blocking reagent prior application of the samples. Standards ranging from 3.9 ng·ml⁻¹ to 250 ng·ml⁻¹, controls and samples were prepared in dilution medium containing 10% FBS and applied in duplicates. After incubation with peroxidase conjugated detection antibody the peroxidase substrate o-Phenylenediamine (Sigma) was used to trigger the specific colorimetric reaction, which was stopped after 10 minutes by addition of 3 M HCl. The read out was done using a microplate spectrophotometer by measurement at 490 nm and 690 nm. The blank was subtracted from all samples and the OD difference between the two wavelengths was used for the determination of sample concentration using a standard curve calculated by 4-parameter fit. Raw data were analysed using either SoftMax Pro, or Tecan software and exported to Excel for further interpretation. The reference standard was chosen according to the protein of interest in the samples. Clone screening approaches were performed as single samples to measure more samples per plate.

8.4.3.3   Isoelectric focussing

Isoelectric focussing (IEF) was used to separate recombinant proteins based on the pI. Proteins are separated in a pH gradient across a polyacrylamide gel and move in an electric field until they lose their net charge at the pH corresponding to the pI. The analysis was performed by the production team using Novex® pH 3-10 IEF Gels (Invitrogen) according to a standard procedure based on the protocol recommended by the supplier.

8.4.3.4   Bicinchoninic acid assay

The BCA assay goes back to Smith [157] and was performed by the production team using the BCA™ Protein Assay Kit (Pierce) according to the suppliers protocol.

8.4.4 Databases and online tools

Table 30: Databases and online tools.

Purpose	Database or online tool
Comparison of protein and nucleotide sequences to databases	Basic Logical Alignment Tool (BLAST) [158]
Prediction of isoelectric point from amino acid sequence	Compute pI/MW [159, 160]
Prediction of signal peptide cleavage site	SignalP [161, 162]
Prediction of N-glycosylation sites	NetNGlyc 1.0 Server [http://www.cbs.dtu.dk/services/NetNGlyc/]
Prediction of mucin type O-glycosylation sites	NetOGlyc 3.1 Server [163] and OGPET v1.0 [http://ogpet.utep.edu/OGPET/]
Prediction of serine, threonine and tyrosine phosphorylation sites	NetPhos 2.0 Server [164]
Prediction of tyrosine sulfation sites	Sulfinator [165]

8.4.5 Formula

$$h(k \mid N, M, n) = \frac{\binom{M}{k} \cdot \binom{N-M}{n-k}}{\binom{N}{n}}$$

Formula 1: Hypergeometric distribution for calculation of probability h that exactly k copies of clone X are being analysed assuming N clones in the pool, M copies of clone X in the pool and n cells being analysed. Calculated with the binominal coefficient described in Formula 2.

$$\binom{a}{b} = \frac{a!}{b! \cdot (a-b)!}$$

Formula 2: Binomial coefficient.

$$efficiency = \frac{c}{w \cdot u} \cdot 100\%$$

Formula 3: Cloning efficiency. Efficiency of limited dilution cloning experiments is described by the percentage of microplate wells with growth per inoculated well. With c: number of clones (wells with cell growth); w: number of wells inoculated; u: average number of cells inoculated per well.

$$CD_{VOL\,j} = \frac{(X_j - X_i)}{\ln\left(X_j / X_i\right)} \cdot (t_j - t_i) + CD_{VOL_i}$$

Formula 4: Cumulative volumetric cell-days. This parameter describes the potential biologic production capacity in a bioreactor and is calculated as described earlier [155] with the modification of using days instead of hours as time unit. With CD_{VOL}: cumulative volumetric cell-days; X: viable cell count; t: culture time; i: initial; j: actual

$$CSP = \frac{(P_j - P_i)}{(CD_{VOL_j} - CD_{VOL_i})}$$

Formula 5: Cell specific productivity. This parameter describes the production capacity of a recombinant cell. With CSP: cell specific productivity; P: product concentration; CD_{VOL}: cumulative volumetric cell-days; i: initial; j: actual.

$$frequency = \frac{c}{w \cdot u}$$

Formula 6: Gene targeting frequency. Efficiency of gene targeting experiments was calculated similar to cloning efficiency. Targeting frequency describes the fraction of targeted clones per transfected cell. With c: number of clones (wells with cell growth); w: number of wells inoculated; u: average number of treated cells inoculated per well.

$$efficiency = \frac{fc}{vc} * 100\%$$

Formula 7: Transfection efficiency. With fc: concentration fluorescent cells and vc: concentration of viable cells.

8.5 Abbreviations

ATCC	American Type Culture Collection
APRT	Adenine phosphoribosyltransferase
BCA	Bicinchoninic acid
BLAST	Basic Logical Alignment Tool
bp	Base pair
CHO	Chinese hamster ovary
CMV	Cytomegalovirus
CRB-15	Cytokine receptor blocker (IL-15 receptor)
DHFR	Dihydrofolate reductase
D-loop	Displacement loop
DMSO	Dimethylsulfoxide
Δneo	Neo lacking a promoter and the initiator codon
DSB	Double-strand break
dsDNA	Double-stranded DNA
EASE	Expression augmenting sequence element
EEO	Electroendoosmosis
ELISA	Enzyme-linked immunosorbent assay
ES cell	Embryonic stem cell
F(ab)2	Fragment antigen binding
FBS	Foetal bovine serum
Fc region	Fragment crystallisable region
G418	Geneticin sulphate
GS	Glutamine synthetase
HEPES	4-(2-hydroxyethyl)-1-piperazineethanesulfonic acid
HPRT	Hypoxanthine-guanine phosphoribosyl transferase
hSEAP	Human secreted alkaline phosphatase
HSV	Herpes simplex virus
Hygro	Hygromycin phosphotransferase
IEF	Isoelectric focussing
IgG	Immunoglobin G
LB	Lysogeny broth
LD	Limited Dilution
$\mu U \cdot c^{-1} \cdot d^{-1}$	micro Unit per cell and day ($\mu U \cdot cell^{-1} \cdot day^{-1}$)

min	Minute
mRNA	messenger RNA
MSX	Methionine sulfoximine
MTX	Methotrexate
Neo	Neomycin phosphotransferase
neoΔ	Neo expression cassette lacking polyadenylation sequences and terminal 161 bp coding sequence
NF	Nucleofection experiment
No.	Number
ori	Origin of replication
PCR	Polymerase chain reaction
PE	R-Phycoerythrin
PEI	Polyethylenimine
$pg{\cdot}c^{-1}{\cdot}d^{-1}$	pg per cell and day ($pg{\cdot}cell^{-1}{\cdot}day^{-1}$)
pUC	Cloning vector produced at University of California
rpm	Revolutions per minute
rRNA	ribosomal RNA
sl	Standard litre
S/MAR	Scaffold- or matrix-attachment region
ssDNA	Single-stranded DNA
STAR element	Stabilising and anti repressor element
tk	Thymidine kinase
TNF	Tumor necrosis factor
UCOE	Ubiquitous chromatin opening element
UV	Ultraviolet
VIS	Visible spectrum of light

8.6 Index of tables

8.7 Index of figures

9 Bibliography

1. Cohen, S.N., Chang, A.C., Boyer, H.W. & Helling, R.B. Construction of biologically functional bacterial plasmids in vitro. *Proc Natl Acad Sci U S A* **70**, 3240-3244 (1973).
2. Walsh, G. Biopharmaceutical benchmarks 2006. *Nat Biotechnol* **24**, 769-776 (2006).
3. Wurm, F.M. Production of recombinant protein therapeutics in cultivated mammalian cells. *Nat Biotechnol* **22**, 1393-1398 (2004).
4. Butler, M. Animal cell cultures: recent achievements and perspectives in the production of biopharmaceuticals. *Appl Microbiol Biotechnol* **68**, 283-291 (2005).
5. DiMasi, J.A., Hansen, R.W. & Grabowski, H.G. The price of innovation: new estimates of drug development costs. *J Health Econ* **22**, 151-185 (2003).
6. Tamimi, N.A. & Ellis, P. Drug development: from concept to marketing! *Nephron Clin Pract* **113**, c125-131 (2009).
7. Kalwy, S., Rance, J. & Young, R. Toward more efficient protein expression: keep the message simple. *Mol Biotechnol* **34**, 151-156 (2006).
8. Kozak, M. Point mutations define a sequence flanking the AUG initiator codon that modulates translation by eukaryotic ribosomes. *Cell* **44**, 283-292 (1986).
9. Kozak, M. An analysis of 5'-noncoding sequences from 699 vertebrate messenger RNAs. *Nucleic Acids Res* **15**, 8125-8148 (1987).
10. Kim, D.W., Uetsuki, T., Kaziro, Y., Yamaguchi, N. & Sugano, S. Use of the human elongation factor 1 alpha promoter as a versatile and efficient expression system. *Gene* **91**, 217-223 (1990).
11. Gopalkrishnan, R.V., Christiansen, K.A., Goldstein, N.I., DePinho, R.A. & Fisher, P.B. Use of the human EF-1alpha promoter for expression can significantly increase success in establishing stable cell lines with consistent expression: a study using the tetracycline-inducible system in human cancer cells. *Nucleic Acids Res* **27**, 4775-4782 (1999).
12. Running Deer, J. & Allison, D.S. High-level expression of proteins in mammalian cells using transcription regulatory sequences from the Chinese hamster EF-1alpha gene. *Biotechnol Prog* **20**, 880-889 (2004).
13. Thomsen, D.R., Stenberg, R.M., Goins, W.F. & Stinski, M.F. Promoter-regulatory region of the major immediate early gene of human cytomegalovirus. *Proc Natl Acad Sci U S A* **81**, 659-663 (1984).
14. Addison, C.L., Hitt, M., Kunsken, D. & Graham, F.L. Comparison of the human versus murine cytomegalovirus immediate early gene promoters for transgene expression by adenoviral vectors. *J Gen Virol* **78 (Pt 7)**, 1653-1661 (1997).
15. Chapman, B.S., Thayer, R.M., Vincent, K.A. & Haigwood, N.L. Effect of intron A from human cytomegalovirus (Towne) immediate-early gene on heterologous expression in mammalian cells. *Nucleic Acids Res* **19**, 3979-3986 (1991).
16. Goodwin, E.C. & Rottman, F.M. The 3'-flanking sequence of the bovine growth hormone gene contains novel elements required for efficient and accurate polyadenylation. *J Biol Chem* **267**, 16330-16334 (1992).

17. Carswell, S. & Alwine, J.C. Efficiency of utilization of the simian virus 40 late polyadenylation site: effects of upstream sequences. *Mol Cell Biol* **9**, 4248-4258 (1989).

18. Southern, P.J. & Berg, P. Transformation of mammalian cells to antibiotic resistance with a bacterial gene under control of the SV40 early region promoter. *J Mol Appl Genet* **1**, 327-341 (1982).

19. Gritz, L. & Davies, J. Plasmid-encoded hygromycin B resistance: the sequence of hygromycin B phosphotransferase gene and its expression in Escherichia coli and Saccharomyces cerevisiae. *Gene* **25**, 179-188 (1983).

20. Lucas, B.K. et al. High-level production of recombinant proteins in CHO cells using a dicistronic DHFR intron expression vector. *Nucleic Acids Res* **24**, 1774-1779 (1996).

21. Sautter, K. & Enenkel, B. Selection of high-producing CHO cells using NPT selection marker with reduced enzyme activity. *Biotechnol Bioeng* **89**, 530-538 (2005).

22. Bianchi, A.A. & McGrew, J.T. High-level expression of full-length antibodies using trans-complementing expression vectors. *Biotechnol Bioeng* **84**, 439-444 (2003).

23. Kaufman, R.J. & Sharp, P.A. Amplification and expression of sequences cotransfected with a modular dihydrofolate reductase complementary dna gene. *J Mol Biol* **159**, 601-621 (1982).

24. Gasser, C.S., Simonsen, C.C., Schilling, J.W. & Schimke, R.T. Expression of abbreviated mouse dihydrofolate reductase genes in cultured hamster cells. *Proc Natl Acad Sci U S A* **79**, 6522-6526 (1982).

25. Young, A.P. & Ringold, G.M. Mouse 3T6 cells that overproduce glutamine synthetase. *J Biol Chem* **258**, 11260-11266 (1983).

26. Urlaub, G. & Chasin, L.A. Isolation of Chinese hamster cell mutants deficient in dihydrofolate reductase activity. *Proc Natl Acad Sci U S A* **77**, 4216-4220 (1980).

27. Urlaub, G., Kas, E., Carothers, A.M. & Chasin, L.A. Deletion of the diploid dihydrofolate reductase locus from cultured mammalian cells. *Cell* **33**, 405-412 (1983).

28. de la Cruz Edmonds, M.C. et al. Development of transfection and high-producer screening protocols for the CHOK1SV cell system. *Mol Biotechnol* **34**, 179-190 (2006).

29. Loc, P.V. & Stratling, W.H. The matrix attachment regions of the chicken lysozyme gene co-map with the boundaries of the chromatin domain. *EMBO J* **7**, 655-664 (1988).

30. Kim, J.M. et al. Improved recombinant gene expression in CHO cells using matrix attachment regions. *J Biotechnol* **107**, 95-105 (2004).

31. Girod, P.A., Zahn-Zabal, M. & Mermod, N. Use of the chicken lysozyme 5' matrix attachment region to generate high producer CHO cell lines. *Biotechnol Bioeng* **91**, 1-11 (2005).

32. Recillas-Targa, F. et al. Position-effect protection and enhancer blocking by the chicken beta-globin insulator are separable activities. *Proc Natl Acad Sci U S A* **99**, 6883-6888 (2002).

33. Kwaks, T.H. et al. Identification of anti-repressor elements that confer high and stable protein production in mammalian cells. *Nat Biotechnol* **21**, 553-558 (2003).

34. Aldrich, T.L., Viaje, A. & Morris, A.E. EASE vectors for rapid stable expression of recombinant antibodies. *Biotechnol Prog* **19**, 1433-1438 (2003).

35. Williams, S. et al. CpG-island fragments from the HNRPA2B1/CBX3 genomic locus reduce silencing and enhance transgene expression from the hCMV promoter/enhancer in mammalian cells. *BMC Biotechnol* **5**, 17 (2005).

36. Jayapal, K.P., Wlaschin, K.F., Hu, W.-S. & Yap, M.G.S. Recombinant Protein Therapeutics from CHO Cells - 20 Years and Counting. *Chemical Engineering Progress* **103**, 40-47 (2007).

37. Graham, F.L. & van der Eb, A.J. A new technique for the assay of infectivity of human adenovirus 5 DNA. *Virology* **52**, 456-467 (1973).

38. Neumann, E., Schaefer-Ridder, M., Wang, Y. & Hofschneider, P.H. Gene transfer into mouse lyoma cells by electroporation in high electric fields. *EMBO J* **1**, 841-845 (1982).

39. Felgner, P.L. et al. Lipofection: a highly efficient, lipid-mediated DNA-transfection procedure. *Proc Natl Acad Sci U S A* **84**, 7413-7417 (1987).

40. Boussif, O. et al. A versatile vector for gene and oligonucleotide transfer into cells in culture and in vivo: polyethylenimine. *Proc Natl Acad Sci U S A* **92**, 7297-7301 (1995).

41. Feng, Y.Q. et al. Site-specific chromosomal integration in mammalian cells: highly efficient CRE recombinase-mediated cassette exchange. *J Mol Biol* **292**, 779-785 (1999).

42. Seth, G., Charaniya, S., Wlaschin, K.F. & Hu, W.S. In pursuit of a super producer-alternative paths to high producing recombinant mammalian cells. *Curr Opin Biotechnol* **18**, 557-564 (2007).

43. Browne, S.M. & Al-Rubeai, M. Selection methods for high-producing mammalian cell lines. *Trends Biotechnol* **25**, 425-432 (2007).

44. Derouazi, M. et al. Genetic characterization of CHO production host DG44 and derivative recombinant cell lines. *Biochem Biophys Res Commun* **340**, 1069-1077 (2006).

45. Jun, S.C., Kim, M.S., Hong, H.J. & Lee, G.M. Limitations to the development of humanized antibody producing Chinese hamster ovary cells using glutamine synthetase-mediated gene amplification. *Biotechnol Prog* **22**, 770-780 (2006).

46. Ozturk, S.S. & Hu, W.-S. Cell culture technology for pharmaceutical and cell-based therapies. (Taylor & Francis, Boca Raton; 2005).

47. Pham, P.L., Kamen, A. & Durocher, Y. Large-scale transfection of mammalian cells for the fast production of recombinant protein. *Mol Biotechnol* **34**, 225-237 (2006).

48. Derouazi, M. et al. Serum-free large-scale transient transfection of CHO cells. *Biotechnol Bioeng* **87**, 537-545 (2004).

49. Haldankar, R., Li, D., Saremi, Z., Baikalov, C. & Deshpande, R. Serum-free suspension large-scale transient transfection of CHO cells in WAVE bioreactors. *Mol Biotechnol* **34**, 191-199 (2006).

50. De Jesus, M.J. et al. TubeSpin satellites: a fast track approach for process development with animal cells using shaking technology. *Biochemical Engineering Journal* **17**, 217-223 (2004).

51. Whitford, W.G. Fed-Batch Mammalian Cell Culture in Bioproduction. *BioProcess International* **4**, 30-40 (2006).

52. Boedeker, B.G. Production processes of licensed recombinant factor VIII preparations. *Semin Thromb Hemost* **27**, 385-394 (2001).

53. Yang, J.D. et al. Fed-batch bioreactor process scale-up from 3-L to 2,500-L scale for monoclonal antibody production from cell culture. *Biotechnol Bioeng* **98**, 141-154 (2007).

54. Smithies, O., Gregg, R.G., Boggs, S.S., Koralewski, M.A. & Kucherlapati, R.S. Insertion of DNA sequences into the human chromosomal beta-globin locus by homologous recombination. *Nature* **317**, 230-234 (1985).

55. Thomas, K.R. & Capecchi, M.R. Site-directed mutagenesis by gene targeting in mouse embryo-derived stem cells. *Cell* **51**, 503-512 (1987).

56. Rong, Y.S. & Golic, K.G. Gene targeting by homologous recombination in Drosophila. *Science* **288**, 2013-2018 (2000).

57. Mansour, S.L., Thomas, K.R. & Capecchi, M.R. Disruption of the proto-oncogene int-2 in mouse embryo-derived stem cells: a general strategy for targeting mutations to non-selectable genes. *Nature* **336**, 348-352 (1988).

58. Mansour, S.L., Goddard, J.M. & Capecchi, M.R. Mice homozygous for a targeted disruption of the proto-oncogene int-2 have developmental defects in the tail and inner ear. *Development* **117**, 13-28 (1993).

59. Detloff, P.J. et al. Deletion and replacement of the mouse adult beta-globin genes by a "plug and socket" repeated targeting strategy. *Mol Cell Biol* **14**, 6936-6943 (1994).

60. Lewis, J., Yang, B., Detloff, P. & Smithies, O. Gene modification via "plug and socket" gene targeting. *J Clin Invest* **97**, 3-5 (1996).

61. Cohen-Tannoudji, M. et al. I-SceI-induced gene replacement at a natural locus in embryonic stem cells. *Mol Cell Biol* **18**, 1444-1448 (1998).

62. Muller, U. Ten years of gene targeting: targeted mouse mutants, from vector design to phenotype analysis. *Mech Dev* **82**, 3-21 (1999).

63. Capecchi, M.R. Altering the genome by homologous recombination. *Science* **244**, 1288-1292 (1989).

64. Adair, G.M. et al. Targeted homologous recombination at the endogenous adenine phosphoribosyltransferase locus in Chinese hamster cells. *Proc Natl Acad Sci U S A* **86**, 4574-4578 (1989).

65. te Riele, H., Maandag, E.R. & Berns, A. Highly efficient gene targeting in embryonic stem cells through homologous recombination with isogenic DNA constructs. *Proc Natl Acad Sci U S A* **89**, 5128-5132 (1992).

66. Hasty, P., Rivera-Perez, J. & Bradley, A. The length of homology required for gene targeting in embryonic stem cells. *Mol Cell Biol* **11**, 5586-5591 (1991).

67. Deng, C. & Capecchi, M.R. Reexamination of gene targeting frequency as a function of the extent of homology between the targeting vector and the target locus. *Mol Cell Biol* **12**, 3365-3371 (1992).

68. Capecchi, M. Gene targeting. How efficient can you get? *Nature* **348**, 109 (1990).

69. Orr-Weaver, T.L., Szostak, J.W. & Rothstein, R.J. Yeast transformation: a model system for the study of recombination. *Proc Natl Acad Sci U S A* **78**, 6354-6358 (1981).

70. Puchta, H., Dujon, B. & Hohn, B. Homologous recombination in plant cells is enhanced by in vivo induction of double strand breaks into DNA by a site-specific endonuclease. *Nucleic Acids Res* **21**, 5034-5040 (1993).

71. Rouet, P., Smih, F. & Jasin, M. Expression of a site-specific endonuclease stimulates homologous recombination in mammalian cells. *Proc Natl Acad Sci U S A* **91**, 6064-6068 (1994).

72. Rouet, P., Smih, F. & Jasin, M. Introduction of double-strand breaks into the genome of mouse cells by expression of a rare-cutting endonuclease. *Mol Cell Biol* **14**, 8096-8106 (1994).

73. Choulika, A., Perrin, A., Dujon, B. & Nicolas, J.F. Induction of homologous recombination in mammalian chromosomes by using the I-SceI system of Saccharomyces cerevisiae. *Mol Cell Biol* **15**, 1968-1973 (1995).

74. Smih, F., Rouet, P., Romanienko, P.J. & Jasin, M. Double-strand breaks at the target locus stimulate gene targeting in embryonic stem cells. *Nucleic Acids Res* **23**, 5012-5019 (1995).

75. Sauer, B. & Henderson, N. Targeted insertion of exogenous DNA into the eukaryotic genome by the Cre recombinase. *New Biol* **2**, 441-449 (1990).

76. O'Gorman, S., Fox, D.T. & Wahl, G.M. Recombinase-mediated gene activation and site-specific integration in mammalian cells. *Science* **251**, 1351-1355 (1991).

77. Kilby, N.J., Snaith, M.R. & Murray, J.A. Site-specific recombinases: tools for genome engineering. *Trends Genet* **9**, 413-421 (1993).

78. Bode, J. et al. The transgeneticist's toolbox: novel methods for the targeted modification of eukaryotic genomes. *Biol Chem* **381**, 801-813 (2000).

79. Jasin, M. Genetic manipulation of genomes with rare-cutting endonucleases. *Trends Genet* **12**, 224-228 (1996).

80. Dujon, B. Sequence of the intron and flanking exons of the mitochondrial 21S rRNA gene of yeast strains having different alleles at the omega and rib-1 loci. *Cell* **20**, 185-197 (1980).

81. Belfort, M. & Roberts, R.J. Homing endonucleases: keeping the house in order. *Nucleic Acids Res* **25**, 3379-3388 (1997).

82. Colleaux, L. et al. Universal code equivalent of a yeast mitochondrial intron reading frame is expressed into E. coli as a specific double strand endonuclease. *Cell* **44**, 521-533 (1986).

83. Monteilhet, C., Perrin, A., Thierry, A., Colleaux, L. & Dujon, B. Purification and characterization of the in vitro activity of I-Sce I, a novel and highly specific endonuclease encoded by a group I intron. *Nucleic Acids Res* **18**, 1407-1413 (1990).

84. Colleaux, L., D'Auriol, L., Galibert, F. & Dujon, B. Recognition and cleavage site of the intron-encoded omega transposase. *Proc Natl Acad Sci U S A* **85**, 6022-6026 (1988).

85. Bos, J.L., Heyting, C., Borst, P., Arnberg, A.C. & Van Bruggen, E.F. An insert in the single gene for the large ribosomal RNA in yeast mitochondrial DNA. *Nature* **275**, 336-338 (1978).

86. Jacquier, A. & Dujon, B. An intron-encoded protein is active in a gene conversion process that spreads an intron into a mitochondrial gene. *Cell* **41**, 383-394 (1985).

87.   Macreadie, I.G., Scott, R.M., Zinn, A.R. & Butow, R.A. Transposition of an intron in yeast mitochondria requires a protein encoded by that intron. *Cell* **41**, 395-402 (1985).

88.   Thermes, V. et al. I-SceI meganuclease mediates highly efficient transgenesis in fish. *Mech Dev* **118**, 91-98 (2002).

89.   Richardson, C. & Jasin, M. Frequent chromosomal translocations induced by DNA double-strand breaks. *Nature* **405**, 697-700 (2000).

90.   Perez, C. et al. Factors affecting double-strand break-induced homologous recombination in mammalian cells. *Biotechniques* **39**, 109-115 (2005).

91.   van Gent, D.C., Hoeijmakers, J.H. & Kanaar, R. Chromosomal stability and the DNA double-stranded break connection. *Nat Rev Genet* **2**, 196-206 (2001).

92.   Helleday, T. Pathways for mitotic homologous recombination in mammalian cells. *Mutat Res* **532**, 103-115 (2003).

93.   Pfeiffer, P., Goedecke, W. & Obe, G. Mechanisms of DNA double-strand break repair and their potential to induce chromosomal aberrations. *Mutagenesis* **15**, 289-302 (2000).

94.   Pastwa, E. & Blasiak, J. Non-homologous DNA end joining. *Acta Biochim Pol* **50**, 891-908 (2003).

95.   Liang, F., Han, M., Romanienko, P.J. & Jasin, M. Homology-directed repair is a major double-strand break repair pathway in mammalian cells. *Proc Natl Acad Sci U S A* **95**, 5172-5177 (1998).

96.   Lin, F.L., Sperle, K. & Sternberg, N. Model for homologous recombination during transfer of DNA into mouse L cells: role for DNA ends in the recombination process. *Mol Cell Biol* **4**, 1020-1034 (1984).

97.   Paques, F. & Haber, J.E. Multiple pathways of recombination induced by double-strand breaks in Saccharomyces cerevisiae. *Microbiol Mol Biol Rev* **63**, 349-404 (1999).

98.   Szostak, J.W., Orr-Weaver, T.L., Rothstein, R.J. & Stahl, F.W. The double-strand-break repair model for recombination. *Cell* **33**, 25-35 (1983).

99.   Valancius, V. & Smithies, O. Double-strand gap repair in a mammalian gene targeting reaction. *Mol Cell Biol* **11**, 4389-4397 (1991).

100.   Voelkel-Meiman, K. & Roeder, G.S. Gene conversion tracts stimulated by HOT1-promoted transcription are long and continuous. *Genetics* **126**, 851-867 (1990).

101.   Flores-Rozas, H. & Kolodner, R.D. Links between replication, recombination and genome instability in eukaryotes. *Trends Biochem Sci* **25**, 196-200 (2000).

102.   Haber, J.E. Recombination: a frank view of exchanges and vice versa. *Curr Opin Cell Biol* **12**, 286-292 (2000).

103.   Liang, F., Romanienko, P.J., Weaver, D.T., Jeggo, P.A. & Jasin, M. Chromosomal double-strand break repair in Ku80-deficient cells. *Proc Natl Acad Sci U S A* **93**, 8929-8933 (1996).

104.   Fukushige, S. & Sauer, B. Genomic targeting with a positive-selection lox integration vector allows highly reproducible gene expression in mammalian cells. *Proc Natl Acad Sci U S A* **89**, 7905-7909 (1992).

105.   Puttini, S. et al. Development of a targeted transgenesis strategy in highly differentiated cells: a powerful tool for functional genomic analysis. *J Biotechnol* **116**, 145-151 (2005).

106. Beck, E., Ludwig, G., Auerswald, E.A., Reiss, B. & Schaller, H. Nucleotide sequence and exact localization of the neomycin phosphotransferase gene from transposon Tn5. *Gene* **19**, 327-336 (1982).

107. Greulich, B. & Herrmann, A. New Perspectives in Production of Biopharmaceuticals & Lead Validation: High Producer Cell Lines in Four Weeks. *Poster presented at the EAPB conference Science to Market, October 7-8, 2008, Hannover, Germany.*

108. Greulich, B., Landauer, K. & Herrmann, A. Cell lines in four weeks with the CEMAX® system. *Poster presented at the 21st ESACT Meeting, Dublin, Ireland, June 7-10, 2007.*

109. Barnes, L.M., Bentley, C.M. & Dickson, A.J. Stability of protein production from recombinant mammalian cells. *Biotechnol Bioeng* **81**, 631-639 (2003).

110. Wilson, C., Bellen, H.J. & Gehring, W.J. Position effects on eukaryotic gene expression. *Annu Rev Cell Biol* **6**, 679-714 (1990).

111. Peabody, D.S. Translation initiation at an ACG triplet in mammalian cells. *J Biol Chem* **262**, 11847-11851 (1987).

112. Touriol, C. et al. Generation of protein isoform diversity by alternative initiation of translation at non-AUG codons. *Biol Cell* **95**, 169-178 (2003).

113. Cabaniols, J.P. 2009).

114. Loyter, A., Scangos, G.A. & Ruddle, F.H. Mechanisms of DNA uptake by mammalian cells: fate of exogenously added DNA monitored by the use of fluorescent dyes. *Proc Natl Acad Sci U S A* **79**, 422-426 (1982).

115. Wurm, F.M. & Petropoulos, C.J. Plasmid integration, amplification and cytogenetics in CHO cells: questions and comments. *Biologicals* **22**, 95-102 (1994).

116. Siegfried, Z. & Cedar, H. DNA methylation: a molecular lock. *Curr Biol* **7**, R305-307 (1997).

117. Lukacsovich, T., Yang, D. & Waldman, A.S. Repair of a specific double-strand break generated within a mammalian chromosome by yeast endonuclease I-SceI. *Nucleic Acids Res* **22**, 5649-5657 (1994).

118. Perrin, A., Buckle, M. & Dujon, B. Asymmetrical recognition and activity of the I-SceI endonuclease on its site and on intron-exon junctions. *EMBO J* **12**, 2939-2947 (1993).

119. Sinacore, M.S. et al. CHO DUKX cell lineages preadapted to growth in serum-free suspension culture enable rapid development of cell culture processes for the manufacture of recombinant proteins. *Biotechnol Bioeng* **52**, 518-528 (1996).

120. Greulich, B., Vol. Diplom Ingenieur (Diploma thesis, University of applied sciences Aachen, Campus Jülich, 2005).

121. Kim, S.H. & Lee, G.M. Development of serum-free medium supplemented with hydrolysates for the production of therapeutic antibodies in CHO cell cultures using design of experiments. *Appl Microbiol Biotechnol* **83**, 639-648 (2009).

122. Heidemann, R. et al. The use of peptones as medium additives for the production of a recombinant therapeutic protein in high density perfusion cultures of mammalian cells. *Cytotechnology* **32**, 157-167 (2000).

123. Birch, J.R. & Racher, A.J. Antibody production. *Adv Drug Deliv Rev* **58**, 671-685 (2006).

124. Chusainow, J. et al. A study of monoclonal antibody-producing CHO cell lines: what makes a stable high producer? *Biotechnol Bioeng* **102**, 1182-1196 (2009).

125. Prentice, H.L., Ehrenfels, B.N. & Sisk, W.P. Improving performance of mammalian cells in fed-batch processes through "bioreactor evolution". *Biotechnol Prog* **23**, 458-464 (2007).

126. Cabaniols, J.P. 2009).

127. Takebe, Y. et al. SR alpha promoter: an efficient and versatile mammalian cDNA expression system composed of the simian virus 40 early promoter and the R-U5 segment of human T-cell leukemia virus type 1 long terminal repeat. *Mol Cell Biol* **8**, 466-472 (1988).

128. Sauer, P.W., Burky, J.E., Wesson, M.C., Sternard, H.D. & Qu, L. A high-yielding, generic fed-batch cell culture process for production of recombinant antibodies. *Biotechnol Bioeng* **67**, 585-597 (2000).

129. Kumar, N., Gammell, P. & Clynes, M. Proliferation control strategies to improve productivity and survival during CHO based production culture : A summary of recent methods employed and the effects of proliferation control in product secreting CHO cell lines. *Cytotechnology* **53**, 33-46 (2007).

130. Hartman, T.E. et al. Derivation and characterization of cholesterol-independent non-GS NS0 cell lines for production of recombinant antibodies. *Biotechnol Bioeng* **96**, 294-306 (2007).

131. Jenkins, N., Parekh, R.B. & James, D.C. Getting the glycosylation right: implications for the biotechnology industry. *Nat Biotechnol* **14**, 975-981 (1996).

132. Zahn, S., Abst, K., Palmen, N., Schindler, S. & Herrmann, A. in Cell Technology for Cell Products; Proceedings of the 19th ESACT Meeting, Harrogate, IK, June 5-8, 2005 765-767 2007).

133. Mauser, C. 2008).

134. Rasmussen, B., Davis, R., Thomas, J. & Reddy, P. Isolation, characterization and recombinant protein expression in Veggie-CHO: A serum-free CHO host cell line. *Cytotechnology* **28**, 31-42 (1998).

135. Pak, S.C.O., Hunt, S.M.N., Bridges, M.W., Sleigh, M.J. & Gray, P.P. Super-CHO - A cell line capable of autocrine growth under fully defined protein-free conditions. *Cytotechnology* **22**, 139-146 (1996).

136. Sunstrom, N.A., Baig, M., Cheng, L., Payet Sugyiono, D. & Gray, P. Recombinant insulin-like growth factor-I (IGF-I) production in Super-CHO results in the expression of IGF-I receptor and IGF binding protein 3. *Cytotechnology* **28**, 91-100 (1998).

137. Enenkel, B., Fieder, J., Otto, R. & Krieg, T. (Boehringer Ingelheim Pharma GmbH & Co. KG, WO patent application 2005019442 A3R4; 2005).

138. Valamehr, B. & Seewoester, T. (Amgen Inc., US patent application 20060115901 A1; 2006).

139. Clarke, J., Thorpe, R. & Davis, J.M. in Basic Cell Culture. (ed. J.M. Davis) (Oxford University Press, 1994).

140. Lattenmayer, C. et al. Characterisation of recombinant CHO cell lines by investigation of protein productivities and genetic parameters. *J Biotechnol* **128**, 716-725 (2007).

141. Yamane-Ohnuki, N. et al. Establishment of FUT8 knockout Chinese hamster ovary cells: an ideal host cell line for producing completely defucosylated antibodies with enhanced antibody-dependent cellular cytotoxicity. *Biotechnol Bioeng* **87**, 614-622 (2004).

142. Umana, P., Jean-Mairet, J., Moudry, R., Amstutz, H. & Bailey, J.E. Engineered glycoforms of an antineuroblastoma IgG1 with optimized antibody-dependent cellular cytotoxic activity. *Nat Biotechnol* **17**, 176-180 (1999).

143. Davies, J. et al. Expression of GnTIII in a recombinant anti-CD20 CHO production cell line: Expression of antibodies with altered glycoforms leads to an increase in ADCC through higher affinity for FC gamma RIII. *Biotechnol Bioeng* **74**, 288-294 (2001).

144. Wurm, F.M. et al. Gene transfer and amplification in CHO cells. Efficient methods for maximizing specific productivity and assessment of genetic consequences. *Ann N Y Acad Sci* **782**, 70-78 (1996).

145. Barnes, L.M., Moy, N. & Dickson, A.J. Phenotypic variation during cloning procedures: analysis of the growth behavior of clonal cell lines. *Biotechnol Bioeng* **94**, 530-537 (2006).

146. Meng, Y.G., Liang, J., Wong, W.L. & Chisholm, V. Green fluorescent protein as a second selectable marker for selection of high producing clones from transfected CHO cells. *Gene* **242**, 201-207 (2000).

147. Mancia, F. et al. Optimization of protein production in mammalian cells with a coexpressed fluorescent marker. *Structure* **12**, 1355-1360 (2004).

148. Gubin, A.N., Reddy, B., Njoroge, J.M. & Miller, J.L. Long-term, stable expression of green fluorescent protein in mammalian cells. *Biochem Biophys Res Commun* **236**, 347-350 (1997).

149. Ramesh, N., Kim, S.T., Wei, M.Q., Khalighi, M. & Osborne, W.R. High-titer bicistronic retroviral vectors employing foot-and-mouth disease virus internal ribosome entry site. *Nucleic Acids Res* **24**, 2697-2700 (1996).

150. Sambrook, J. & Russell, D.W. Molecular cloning : a laboratory manual, Edn. 3rd. (Cold Spring Harbor Laboratory Press, Cold Spring Harbor, N.Y.; 2001).

151. Mullis, K.B. & Faloona, F.A. Specific synthesis of DNA in vitro via a polymerase-catalyzed chain reaction. *Methods Enzymol* **155**, 335-350 (1987).

152. Rychlik, W. & Rhoads, R.E. A computer program for choosing optimal oligonucleotides for filter hybridization, sequencing and in vitro amplification of DNA. *Nucleic Acids Res* **17**, 8543-8551 (1989).

153. Birnboim, H.C. & Doly, J. A rapid alkaline extraction procedure for screening recombinant plasmid DNA. *Nucleic Acids Res* **7**, 1513-1523 (1979).

154. Southern, E.M. Detection of specific sequences among DNA fragments separated by gel electrophoresis. *J Mol Biol* **98**, 503-517 (1975).

155. Dutton, R., Scharer, J. & Moo-Young, M. Descriptive parameter evaluation in mammalian cell culture. *Cytotechnology* **26**, 139-152 (1998).

156. Cullen, B.R. Utility of the secreted placental alkaline phosphatase reporter enzyme. *Methods Enzymol* **326**, 159-164 (2000).

157. Smith, P.K. et al. Measurement of protein using bicinchoninic acid. *Anal Biochem* **150**, 76-85 (1985).

158. Altschul, S.F. et al. Gapped BLAST and PSI-BLAST: a new generation of protein database search programs. *Nucleic Acids Res* **25**, 3389-3402 (1997).

159. Bjellqvist, B. et al. The focusing positions of polypeptides in immobilized pH gradients can be predicted from their amino acid sequences. *Electrophoresis* **14**, 1023-1031 (1993).

160. Bjellqvist, B., Basse, B., Olsen, E. & Celis, J.E. Reference points for comparisons of two-dimensional maps of proteins from different human cell types defined in a pH scale where isoelectric points correlate with polypeptide compositions. *Electrophoresis* **15**, 529-539 (1994).

161. Nielsen, H., Engelbrecht, J., Brunak, S. & von Heijne, G. Identification of prokaryotic and eukaryotic signal peptides and prediction of their cleavage sites. *Protein Eng* **10**, 1-6 (1997).

162. Bendtsen, J.D., Nielsen, H., von Heijne, G. & Brunak, S. Improved prediction of signal peptides: SignalP 3.0. *J Mol Biol* **340**, 783-795 (2004).

163. Julenius, K., Molgaard, A., Gupta, R. & Brunak, S. Prediction, conservation analysis, and structural characterization of mammalian mucin-type O-glycosylation sites. *Glycobiology* **15**, 153-164 (2005).

164. Blom, N., Gammeltoft, S. & Brunak, S. Sequence and structure-based prediction of eukaryotic protein phosphorylation sites. *J Mol Biol* **294**, 1351-1362 (1999).

165. Monigatti, F., Gasteiger, E., Bairoch, A. & Jung, E. The Sulfinator: predicting tyrosine sulfation sites in protein sequences. *Bioinformatics* **18**, 769-770 (2002).

Publications

Poster and Proceedings

Greulich, B., Abts, H. F., Zahn, S., and Herrmann, A.: A New Host Cell-line for the Fast, Reproducible and Efficient Production of Biopharmaceuticals. In Cell Technology for Cell Products, 2007; Proceedings of the 19th ESACT Meeting, Harrogate, UK, June 5-8, 2005; p769-771

Greulich, B., Urbschat, A., Abts, H. F., and Herrmann, A.: Flexible generation of high yield-production cell lines using modified CHO-cells and double-strand break induced homologous recombination. Poster presented at Bioperspectives 2007, May 30-April 1, Cologne, Germany

Greulich, B., Urbschat, A., Herrmann, A., and Abts, H. F.: Reproducible generation of high-yield production cell lines using a wildcard-cell system based on double strand break induced homologous recombination for site specific integration. Poster presented at the 20th ESACT Meeting, Dresden, Germany, June 17-20, 2007

Greulich, B., and Herrmann, A.: New Perspectives in Production of Biopharmaceuticals & Lead Validation: High Producer Cell Lines in Four Weeks. Poster presented at the EAPB conference Science to Market, October 7-8, 2008, Hannover, Germany

Greulich, B., Landauer, K., and Herrmann, A.: Cell lines in four weeks with the CEMAX® system. Poster presented at the 21st ESACT Meeting, Dublin, Ireland, June 7-10, 2009. Manuscript accepted for publishing.

Patent

Herrmann, A., Abts, H. F., Greulich, B.: Methods and materials for the reproducible generation of high producer cell lines for recombinant proteins. 2009, Celonic AG, international patent applications WO 2009118192 A1 and EP 2105505 A1.